CHONDRICHTHYES - MULTIDISCIPLINARY APPROACH

Edited by **Luis Fernando da Silva Rodrigues Filho** and **João Bráullio de Luna Sales**

Chondrichthyes - Multidisciplinary Approach

http://dx.doi.org/10.5772/65879
Edited by Luis Fernando da Silva Rodrigues Filho and João Bráullio de Luna Sales

Contributors

Sergio Vitale, Erich Ritter, Raid Amin, Sara Bonanomi, Alessandro Colombelli, Jure Brčić, Jacopo Pulcinella, Emilio Notti, Antonello Sala, Divino Bruno Da Cunha, Luis Fernando Rodrigues-Filho, João Bráullio De Luna Sales, Getulio Rincon, Rodrigo Mazzoleni, Ana Rita Palmeira, Rosangela Lessa

Published by InTech
Janeza Trdine 9, 51000 Rijeka, Croatia

Publishing Process Manager Edi Lipovic
Technical Editor SPi Global
Cover InTech Design team

First published December, 2017
Printed in Croatia
Legal deposit, Croatia: National and University Library in Zagreb

Additional hard copies can be obtained from orders@intechopen.com

Chondrichthyes - Multidisciplinary Approach , Edited by Luis Fernando da Silva Rodrigues Filho and João Bráullio de Luna Sales
p. cm.
Print ISBN 978-953-51-3711-5
Online ISBN 978-953-51-3712-2

For EU product safety concerns:
IN TECH d.o.o., Prolaz Marije Krucifikse Kozulić 3, 51000 Rijeka, Croatia,
info@intechopen.com or visit our website at intechopen.com.

Meet the editors

Prof. Dr Luis Fernando da Silva Rodrigues Filho graduated in Biology from the Federal University of Para, Bragança Campus, and acquired his master's and PhD degrees in Environmental Biology from the same university. He is currently a professor at the Federal Rural University of Amazonia (UFRA), Capanema Campus; a researcher at the Coastal Ecosystem Institute (IECOS) at the Federal University of Pará (UFPa); and co-supervisor of Postgraduate Program in Environmental Biology at the Federal University of Pará, Campus of Bragança. He is also a reviewer of numerous national and international scientific journals. His research interests include animal genetics: phylogeny, phylogeography, and genetics of population in marine species.

Prof. Dr. João Bráullio de Luna Sales graduated in Biology from the Federal University of Para Campus of Bragança, with master's degree in Marine Biology and PhD degree in Biological Resources from the same institution. Currently, he is an associate professor at the Federal University of Pará of Marajó-Brevis Campus. He is also an associate supervisor of Post-Graduate Program in Environmental Biology at the Federal University of Para, Campus of Bragança, and a reviewer of numerous international journals. His research mainly focuses on animal genetics with emphasis on phylogeny, biogeography, and population genetics, principally in marine and freshwater species.

Contents

Preface

Chondrichthyes are a group of cartilaginous fish, where we have sharks, rays, and chimeras as members. This group plays an important role in aquatic ecology by acting as top predators in the food chain. However, many populations of Chondrichthyes are threatened by various factors (increased direct fishing, high mortality rate as accompanying fauna, marine pollution, habitat destruction, etc.).

The use of the Chondrichthyes as a fishing resource already happens to years, but its capture was of accidental way. In the last two decades, this scenario has given way to a directed fishing scenario due to its commercial valorization, mainly of its by-products. As a result, shark and ray fishing reached alarming rates, leading to 25% of known species to be listed as threatened with extinction. Added to this comes the fact that these species are strategists, which is determinant for the vulnerability of this taxon to the fish, since they usually present a low production of puppies and a late maturity.

These declines are evident in many parts of the world and have come to the attention of scientists, conservation organizations, the media, and the general public. Fisheries legislation regulating international fisheries markets has been amended to provide greater protection for this group along with other species of fish. However, little is known about these species, which reinforces the importance of studies in order to have a better understanding of these stocks, as well as to identify the influences of anthropic action through fishing.

In response to knowledge on the low sustainability of cartilaginous fish fisheries on a global scale, FAO has developed an international plan of action for the management and conservation of these fish, with the aim of developing and implementing national plans of action, ensuring management and conservation of these stocks, having as main recommendation the collection of information about the Chondrichthyes, especially the sharks. Even so, this group is still little known in terms of biodiversity, ecology, behavior, and a host of other characteristics relevant to this taxon, which is very worrying.

Chondrichthyes - Multidisciplinary Approach attempts to portray, through an assembly of up-to-date information on cartilaginous fish, a general and current view of this taxon. Thus, "Chondrichthyes - Multidisciplinary Approach" is structured in two sections ("Chondrichthyes, Men and Fishing" and "Biodiversity, Phylogenetic Systematics in Chondrichthyes"), where we discuss relevant topics; the various interactions of man with this group, including fishing; and other aspects, as species diversity and their systematic relationships.

Then a survey on this group will serve as an indispensable source of access to more accurate and detailed information on the Chondrichthyes. This will facilitate the preparation of new research and even more sustainable exploitation plans for this fishing resource, which are still very difficult due to a lack of knowledge.

Prof. Dr. Luis Fernando da Silva Rodrigues Filho
Federal Rural University of Amazonia, Capanema Campus,
Undergraduate Course in Biological Sciences, Pará, Brazil

Prof. Dr. João Bráullio de Luna Sales
Federal University of Pará, Marajó-Breves Campus,
Faculty of Natural Sciences, Pará, Brazil

Federal University of Pará, Belém Campus,
Laboratory of Integrated Lepidopterology, Pará, Brazil

Section 1

Chondrichthyes, Men and Fishing

Chapter 1

Introductory Chapter: The Elasmobranchs as a Fishery Resource

Luis Fernando da Silva Rodrigues Filho and
João Bráullio de Luna Sales

Additional information is available at the end of the chapter

http://dx.doi.org/10.5772/intechopen.71355

1. Introduction

The elasmobranchs (sharks, rays, and skates) represent one of the most ancient vertebrate lineages, which have survived a number of different mass extinction events over the past 400 million years [1]. Approximately 1200 elasmobranch species are currently found in the world's oceans [2]. These fishes have a cartilaginous skeleton, with five to seven branchial clefts [3, 4].

However, the full diversity of the elasmobranchs is still unclear, and a number of species have been described in recent years, principally from the waters of Asia and Oceania [5–8]. In one of the most recent taxonomic reviews, the sharks were defined as a group with 8 orders, 34 families, and approximately 500 species. The rays include 5 orders, 20 families, and around 574 species [9, 10]. In recent years, the application of molecular tools in combination with morphological analyses has led to a significant increase in the known diversity of elasmobranch species [6, 11]. One of the most important molecular tools is the DNA barcode [12], which provides a rapid and reliable approach for the identification of taxonomic groups, especially those which may contain cryptic species, that is, taxa that are morphologically indistinguishable, but genetically distinct, highlighting the need for systematic reviews [7, 13, 14].

Recent studies of elasmobranch diversity have focused primarily on the Asian domain and have included the description of new species [5, 7, 15], the re-organization of some families [11], and the re-establishment of others [8]. All these diversities have been "discovered," thanks to the systematic combination of molecular and morphological approaches, with molecular tools being especially useful for the delimitation of groups or species thought to contain cryptic diversity [16, 17] or that need internal review [18]. Morphological analyses, in turn, provide the basic parameters necessary for the identification and description of taxa [8].

This introductory chapter highlights the gaps in our understanding of the diversity of the Chondrichthyes and provides important background for the understanding of the central issues that will be discussed in this book.

2. Elasmobranchs and fisheries

The trade in fishery resources, such as elasmobranchs, has long been an important economic activity of coastal populations, which exploit not only the meat of these fishes but also a number of subproducts derived from these animals [13, 14, 19, 20]. Worldwide, however, the exploitation of this resource has increased at an alarming rate in recent years and often involves unsustainable practices, which have led many species to the brink of extinction [21]. Overfishing is a major cause of the decline of fish stocks around the world, and the populations of sharks and rays have declined drastically in many regions [22, 23].

Industrial fisheries are estimated to have reduced the biomass of large fish by approximately 10% [24]. As a result of overfishing, the stocks of many elasmobranch species have declined steadily around the world, and they are now at risk of extinction. A number of factors have contributed to this decline, including the degradation of habitats essential to the development of the species and bycatch [25–29].

The reduction in the availability of other fishery resources has also led to an increase in the consumption of elasmobranchs. While the meat of sharks and rays demands a lower market price than that of most other marine fish species, there is an enormous market for shark fin, which is highly valued [30, 31]. The high price of shark fin stimulates the practice of finning, in which the fins are extracted and the shark carcass is discarded. The thriving Asian market for shark fin stimulates high prices for the byproduct and favors predatory fishery practices [27, 28, 32, 33].

In a recent global study of fishery catches of the Chondrichthyes, [34] demonstrated that these fishery practices, combined with the increasing demand for byproducts, have resulted in a decline in stocks of approximately 20% over the past decade, highlighting the association between fishing pressure and impacts on the ecosystem. In addition to predatory fishery practices, the biological characteristics of the elasmobranchs, which are *K* strategists, underpin the vulnerability of this taxon to fisheries. Species that are *K* strategists typically have a low reproductive output and delayed sexual maturation, which contribute to the potential risks of fishery exploitation [35].

3. Biological data vs. fisheries

The internal reproductive system is one of the characteristics that have contributed to the evolutionary success of the Chondrichthyes [36]. However, the reproductive characteristics of this group also result in low rates of population expansion, extended longevity, slow growth, late sexual maturation, and the production of relatively few offspring. Together, these characteristics represent a high level of investment on the part of the female in the reproductive process and in each individual offspring, resulting in a progeny relatively small in number but high in quality, with an extended life expectancy, which may compensate for natural rates of mortality [3, 4, 37].

The ongoing growth in the commercial exploitation of the elasmobranchs has reinforced the need for more adequate public policies and management strategies designed to guarantee the sustainability of this fishery resource. The development of measures of this type is increasingly difficult, given the general lack of data, not only on the biology of most shark and ray species but also on the fisheries that target these species [38]. Given this, understanding the biological and demographic features of the target species, as well as the effects of fisheries on their populations, will be fundamentally important for the development of adequate management programs [39].

Perhaps, the most interesting aspect of the life history strategy of the elasmobranchs is their reproductive biology, which is geared to the production of small progenies. In most species, progenies contain between 2 and 20 offspring [39–41]. Breeding is not always annual, given that the gestation period of some species may exceed 1 year, while in others, there is an extended pause between gestations [42–44].

These characteristics of the reproductive biology of the chondrichthyans, such as the small size of the progeny, impose both biological (abundance) and economic limitations on the potential for the exploitation of this fishery resource [37]. The late maturation of many species, which may take decades to reach adulthood, and the long gestation periods, of up to 2 years in some cases, further compound these limitations [45].

These reproductive traits characterize the elasmobranchs as typical *K* strategists [46], which implies a series of problems for the exploitation of this group as a fishery resource, given that one principal characteristic of *K* strategists is their dependence on relatively stable habitats and resources, including productive feeding grounds, to guarantee their reproductive success over the long term [39]. These characteristics of *K* strategists hamper considerably the maintenance of elasmobranch stocks in the context of human interference. Urban and industrial processes in the coastal zone have an intense impact on the habitats of many elasmobranch species. In addition, fisheries compete increasingly with sharks for the same resources (prey fish) and are harvesting growing numbers of elasmobranchs, either as target species or as bycatch [47].

As mentioned above, the biological characteristics of the group make the cartilaginous fish highly vulnerable to exploitation by fisheries [25]. In fact, overfishing has been identified as the principal threat to this group [42, 48] and many stocks have declined abruptly in recent years, with some populations being reduced by more than 90% [42, 48, 49]. Estimates of mortality indicate that intense fishery pressure will lead large sharks and many other species with similar biological characteristics to extinction within the near future [50]. In this sense, the elasmobranchs can be considered to be one of the planet's most endangered groups of vertebrates [27].

Given this, research initiatives that aim to contribute to the sustainability of marine species are hampered by the lack of data on the biology of the species and the status of their stocks [29]. In the specific case of sharks and rays, the reliable assessment of stocks is hampered by severe deficiencies in the fishery statistics [51], in particular, the assignment of captured specimens to general groups rather than specific species [52]. The adequate assessment of stocks and the development of effective management measures and even the conservation of species will require more precise data on the diversity of the species harvested and landed [29]. This will permit the development of appropriate conservation measures to guarantee the sustainability of populations over the long term [29, 53].

Many of the chondrichthyan species being exploited by commercial fisheries are not reported in official catch statistics, due in particular to the difficulties of identifying species reliably [54]. One of the principal factors that impede the reliable identification of species is the fact that most individuals are landed as headless carcasses with no tail or fins [53] and thus lack the morphological characteristics necessary for the identification of the species [55].

These problems are exacerbated by the lack of data on the diversity of the elasmobranch fauna of many regions of the world [56, 57]. In recent years, there have been considerable efforts in the Asian region, with wide-ranging phylogenetic and morphological inferences [5, 6], as well as the description of new species [7, 11, 15] and the reinstatement of a number of families [8].

An important step in the understanding of elasmobranch diversity is the identification of species. One of the principal difficulties for the identification of cartilaginous fish is the fact that most specimens are landed in processed form, such as the fins only, for example, which impedes the identification of the species based on morphological traits [4]. Given this, molecular markers are widely used for the identification of species, as in many other types of organisms of economic interest [58, 59].

In fact, molecular tools have become an essential approach for the identification of species, providing a systematic means of circumventing the difficulties imposed by the lack of diagnostic morphological traits [60]. Molecular identification is now used widely in research on elasmobranchs, based on a variety of genetic techniques [14, 32, 60–64]. The most widely used approach, which is both simple and effective and provides a good diagnosis of most animal taxa, is DNA barcoding [12], which is based on the analysis of a small fragment of the mitochondrial *Cytochrome Oxidase subunit I* (COI) gene of approximately 648 nucleotides. This diagnostic tool was developed with the ultimate objective of storing information on all known organisms in a universal database (the BOLD platform—http://www.boldsystems.org/) to be used for the identification of species [12].

The molecular identification of the different species of hammerhead shark (genus *Sphyrna*) provides an excellent example of the utility of these molecular tools for the conservation and management of these organisms [65]. In these sharks, the head is the principal diagnostic trait for the morphological determination of species, although it is invariably removed prior to the landing of catches, impeding the reliable identification of specimens encountered in fish markets. This problem can be overcome by using molecular tools for the identification of specimens, providing an important mechanism for the monitoring of the shark trade, as well as contributing to the development of management plans for the stocks of the different species.

Overall, then, the application of molecular tools for the identification of shark species and the analysis of their genetic diversity will be fundamentally important for the development of effective management plans for the stocks of the different shark species exploited by fisheries in northern Brazil. The genetic monitoring of fishery stocks is seen as a fundamental prerequisite for the sustainability of harvests, as well as the re-establishment of depleted stocks [14].

4. Conclusions

Predatory fishing practices have been fundamental to the inclusion of a large number of chondrichthyan species in the red list of endangered species. The conservation of the Class Chondrichthyes, which includes all cartilaginous fish (sharks, rays and chimeras), has provoked increasing levels of concern in recent years. Fisheries that target elasmobranchs have also increased their activities in recent years, impacting the stocks of these fish on a global scale. Despite this, the diversity, ecology, behavior, and many other characteristics of these species remain virtually unknown, which is a major cause for concern. This chapter provides researchers and other readers interested in this group of fish, not only with information on the Chondrichthyes but also an overview of the current status of this group and in particular the biological characteristics of these fish and their role as a fishery resource. A review of these features, combined with new research on the chondrichthyans, compiled in the chapters of this book, provide an essential source of detailed information on this important vertebrate group.

Author details

Luis Fernando da Silva Rodrigues Filho[1]* and João Bráullio de Luna Sales[2,3]

*Address all correspondence to: lfsrf@yahoo.com.br

1 Undergraduate Course in Biological Sciences, Federal Rural University of Amazonia, Capanema, Pará, Brazil

2 Faculty of Natural Sciences, Federal University of Pará, Breves, Pará, Brazil

3 Laboratory of Lepidopterology and Integrated Ichthyology, Federal University of Pará, Belém, Pará, Brazil

References

[1] Pavan-Kumar A, Gireesh-Babu P, Suresh Babu et al. Molecular phylogeny of elasmobranchs inferred from mitochondrial and molecular markers. Mol. Biol. Rep. 2015;**41**:447-457 doi:10.1007/s11033-013-2879-6

[2] Fowler SL, Cavanagh RD, Camhi M, Burgess GH, Callitet GM et al. Sharks, Rays and Chimaeras: the status of the Chondrichthyan fishes. Status survey. Gland: IUCN/SSC Shark Specialist Group. 2005. 189 p

[3] Franco BA. Identificação das raias-viola Rhinobatus horkelli, Rhinobatus percellens e Zapteryx brevirostris (Chondricthyes, Rhinobatidae) a costa central e sul do Brasil utilizando marcadores moleculares [These]. Botucatu Obispo: Universidade Estadual Julio de Paulista (UNESP); 2010. 62 p

[4] Vooren CM, Klippel S. Diretrizes para conservação de elasmobrânquios ameaçados. In VOOREN,C. M. & KLIPPEL,S. (Eds.), Ações para a conservaçao de tubarões e raias no sul do Brasil (Igaré, pp. 213-228). Porto Alegre. 2005

[5] Last PR, Henderson AC, Naylor GJP. Acroteriobatus omanesis (Batoidea: Rhinobatidae), a new guitarfish from the Gulf of Oman. Zootaxa. 2016a;**4144**(2):276-286. DOI: http://dx.doi.org/10.11646/zootaxa.4144.2.9

[6] Lim KC, Lim P-E, Chong VC, Loh K-H. Molecular and Morphological Analyses Reveals Phylogenetic Relationships of Stingrays Focusing on the Family Dasyatidae (Myliobatiformes). PLoS ONE. 2015;**10**(4):e0120518. https://doi.org/10.1371/journal.pone.0120518

[7] White WT, Furumitsu K, Yamaguchi A. A new Species of Eagle Ray Aetobatus narutobiei from the Northwest Pacific: An Example of the Critical Role Taxonomy Plays in Fisheries and Ecological Science. PLoS ONE. 2013;**8**(12):e83785. https://doi.org/10.1371/journal.pone.0083785

[8] White WT, Naylor GJP. Resurrection of the family Aetobatidae (Myliobatiformes) for the pelagic eagle rays, genus Aetobatus. Zootaxa. 2016;**4139**(3):435-438. DOI: http://dx.doi.org/10.11646/zootaxa.4139.3.10

[9] Ebert DA, Compagno LJV. Biodiversity and systematics of skates (Chondrichthyes: Rjiformes: Rajoidei). Enviromentl Biology of Fishes, 2007;**80**:111-124. DOI 10.1007/s10641-007-9247-0

[10] Compagno LJV. Checklist of living elasmobranchs. In: Hamlett, W.C. (ed.) Sharks, Skates, and Rays. The biology of Elasmobranch Fishes. The Hopkins University Press, Baltimore: 1999; p. 470-498

[11] Last PR, Naylor GJP, Manjaji-Matsumoto M. A revised classification of the family Dasyatidae (Chondrychthyes: Myliobatiformes) based on new morphological and molecular insights. Zootaxa. 2016b;**4139**(3):345-368. DOI: http://dx.doi.org/10.11646/zootaxa.4139.3.2

[12] Hebert PDN. Ratnasingham S, de Waard JR. Barcoding animal life: cytochrome c oxidase subunit 1 divergences among closely related species. Proc. R. Soc. Lond. 2003;**B270**: S596-S599. DOI: 10.1098/rsbl.2003.0025

[13] Cerutti-Pereyra F, Meekan MG, Wei N-WV, O'shea O, Bradshaw CJA, Austin CM. Identification of Rays through DNA Barcoding: An Application for Ecologists. PLoS ONE. 2012;**7**(6):e36479. DOI: 10.1371/journal.pone.0036479

[14] Ward RD, Holmes BH, White WT, Last PR. DNA barcoding Australasian chondrichthyans: results and possible uses in conservation. Mar. Freshwater Res. 2008;**59**:57-71. https://doi.org/10.1071/MF07148

[15] Last PR, Kyne PM, Compagno LJV. A new species of wedgefish Rhynchobatus cooki (Rhinopristiformes, Rhinidae) from the Indo-West Pacific. Zootaxa. 2016c;**4139**(2):233-247. DOI: http://dx.doi.org/10.11646/zootaxa.4139.2.7

[16] Richards VP, Henning M, Witzell W, Shivji MS. Species Delineation and Evolutionary History of the Globally Distributed Spotted Eagle Ray (Aetobatus narinari). Journal of Heredity. 2009;**100**(3):273-283. DOI: https://doi.org/10.1093/jhered/esp005

[17] Schluessel V, Broderick D, Collin SP, Ovorden JR. Evidence for extensive population structure in the White-spotted eagle ray within the Indo-Pacific inferred from mitochondrial gene sequences. Journal of Zoology. 2010;**281**:46-55. DOI: 10.1111/j.1469-7998.2009.00680.x

[18] White WT. A revised generic arrangement for the eagle ray family Myliobatidae, with definitions for the valid genera. Zootaxa. 2014;**3860**(2):149-166. DOI: http://dx.doi.org/10.11646/zootaxa.3860.2.3

[19] Castro J, Pardo BG, Sánchez L, Martínez P. rDNA RFLPs as genetic markers for resource management in brown trout. J Fish Biol. 1999;**55**:221-225. DOI: 10.1111/j.1095-8649.1999.tb00671.x

[20] Séret B. Guia de identificação das principais espécies de raias e tubarões do Atlântico oriental tropical, para observadores de pesca e biólogos. Tradução Portuguese Rui COELHO IUCN Union mondiale pour la Nature. 2006. 75 p

[21] Fisheries and aquaculture software. FishStatJ - software for fishery statistical time series. In: FAO Fisheries and Aquaculture Department [online]. Rome. Updated 21 July 2016. [Cited 12 November 2017]. http://www.fao.org/fishery/statistics/software/fishstatj/en

[22] Baum JK, Myers RA, Kehler, DG, Worm B, Harley SJ, Doherty PA. Collapse and conservation of shark populations in the northwest Atlantic. Science. 2003;**299**:389-392. DOI: 10.1126/science.1079777

[23] Fonteles-Filho AA. Oceanografia, biologia e dinâmica populacional de recursos pesqueiros. Expressão Gráfica Editora, Fortaleza; 2011. 464 p

[24] Myers RA, Worm B. Rapid Worldwide Depletion Of Predatory Fish Communities. Nature. 2003;**423**:280-283

[25] Camhi M, Fowler S, Musick J, Bräutigam A, Fordham S. Sharks and their relatives. Ecology and Conservation. Species Survival Commission, World Conservation Union, Gland, Switzerland. Occasional Paper of the IUCN Species Survival Commission. 1998. 20 p

[26] Diaz-Neto. Proposta de plano de gestão para o uso sustentável de elasmobrânquios sobre-explotados ou ameaçados de sobre-explotação no Brasil. Brasília. 2015

[27] Lucifora LO, García VB, Worm B. Global Diversity Hotspots And Conservation Priorities For Sharks. Plos One. 2011;**6**(5):E19356. https://doi.org/10.1371/journal.pone.0019356

[28] Myers RA, Baum JK, Shepherd TD, Powers SP, Peterson CH. Cascading Effects Of The Loss Of Apex Predatory Sharks From A Coastal Acean. Science. 2007;**315**:1846-1850. DOI: 10.1126/science.1138657

[29] SBEEL - Sociedade Brasileira para o Estudo de Elasmobrânquios. 2005. Plano Nacional de Ação para a Conservação e o Manejo dos estoques de Peixes Elasmobrânquios no Brasil. Recife – PE. Available from: www.sbeel.org.br/downloads/plano.pdf. [Acessed: 2005_05_25]. 99 p

[30] Kotas JE, Petrere MJr, Azevedo VG, Dos Santos S. A pesca de emalhe e de espinhel de superfície na Região Sudeste- Sul do Brasil. Série documentos Revizee – Score Sul, 2005; 72 p

[31] Szpilman M. Tubarões no Brasil: Guia prático de identi cação. Rio de Janeiro: Aqualittera e Mauad, 2004; 160 p

[32] Holmes B, Steinke D, Ward R. Identication of shark and ray ns using DNA barcoding. Fisheries Research. 2009;**95**(2-3):280-288. https://doi.org/10.1016/j.fishres.2008.09.036

[33] Ussami LHF. Análise da variabilidade e estruturação genética do tubarão azul, Prionace glauca (Chondrichthyes, Carcharhinidae) na costa brasileira, utilizando marcadores microssatélites. [These]. Botucatu: Universidade Estadual Paulista. 2011. 66 p

[34] Davidson LNK, Krawchuk MA, Dulvy NK. Why have global shark and ray landings declined: improved management or overfishing? FISH and FISHERIES. 2015. DOI: 10.1111/faf.12119

[35] Bornatowski H, Abilhoa V. Tubarões e raias capturados pela pesca artesanal no Paraná: guia de identificação. Curitiba, Hori Consultoria Ambiental. Hori Cadernos Técnicos. 2012;**4**:124

[36] Castro JL. Sharks of the North American Waters. Texas A & M University Press, Coil. Sm. First edition; (1983). 180 p

[37] Hoenig JM, Gruber SH. Life-history patterns in the elasmobranchs: implications of fisheries management. In: Pratt-Jr, H. L. R. horkelli du plateau continental du Brésil. These de Doctorat de 3 eme cycle. Univ. Bretagne Occidentale. 1990. 250 p

[38] Furtado-Neto MAA, Barros JúNior FVP. Análise da produção pesqueira de elasmobrânquios no estado do Ceará, Brasil, de 1991 a 2003. Arquivos de Ciências do Mar. 2006;**39**:110-116

[39] Correa JPS. Pesca comercial de tubarões e raias em Portugal [These] Aveiro Obispo: Universidade de Aveiro; 2009

[40] Chang WB, Leu MY, Fang LS Embryos of the whale shark Rhincodon typus: early growth and size distribution. Copeia. 1997;(2):444-446. DOI: 10.2307/1447769

[41] Joung SJ, Chen CT, Clark E, Uchida S, Huang WYP. The whale shark, Rhincodon typus, is a livebearer: 300 embryos found in one 'megamamma' supreme. Env. Biol. Fish. 1996;**46**:219-223

[42] Bornatowski H, Braga, RR, Vitule JRS. Shark Mislabeling Threatens Biodiversity. Science. 2013;**340**(6235):923. DOI: 10.1126/science.340.6135.923-a

[43] Branstetter S. Early life-history implications of selected carcharhinoid and lamnoid sharks of the northwest Atlantic. In: H.L. Pratt Jr., S.H. Gruber, and T. Taniuchi (eds.). Elasmobranchs as living resources: advances in the biology, ecology, systematics, and the status of fisheries. NOAA Technical Report; 1990. p. 17-28

[44] Pratt Jr HL, Casey JG. Shark reproductive strategies as a limiting factor in directed fisheries, with a review of Holden's method of estimating growth-parameters. In: H.L. Pratt Jr., S.H. Gruber, and T. Taniuchi (eds.). Elasmobranchs as living resources: advances in the biology, ecology, systematics, and the status of fisheries. NOAA Technical Report. 1990. p. 97-109

[45] Natanson LJ, Casey JG, Kohler NE. Age and growth estimates for the dusky shark, Carcharhinus obscurus, in the western North Atlantic Ocean. Fish. Bull. 1995;**93**:116-126

[46] Gruber SH. Discovering sharks. American Littoral Society, Highlands; 1990. 122 p

[47] Hazin FHV, Zagaglia JR, Geber FO, Wanderley JAMJr, Mattos SMG. (abstract) In: A shark attack outbreak off Recife-PE. American Elasmobranch Society 12th Annual Meeting. Brazil: New Orleans; 2005

[48] Worm B, Davis B, Kettemer L, Ward-Paige CA, Champman D, Heithaus M, et al. Global catches, exploitation rates, and rebuilding options for sharks. Mar Policy. 2013;**40**:194-204. DOI: 10.1016/j.marpol.2012.12.034

[49] Pauly D, Hilborn R, Branch TA. Fisheries: Does catch reflect abundance? Nature. 2013; **494**:303-306. DOI:10.1038/494303a

[50] Myers RA, Worm B. Extinction, survival or recovery of large predatory fishes. Philosophical Transactions of the Royal Society B. 2005;**360**:13-20. DOI: 10.1098/rstb.2004.1573

[51] Rosa RS, Lima FCT. Peixes. In: A. B. M. Machado, G. M. Drummond, et al. Livro vermelho da fauna brasileira ameaçada de extinção. Brasília, DF: MMA, 2008;**2**:1420

[52] Paesch L, Meneses P. Estudios Realizados Sobre Los Elasmobrânquios Dentro Del Rio De La Plata Y La Zona Comum De Pesca Argentino – Uruguaya En El Marco Del —Plan De Investigacion Pesquera‖. Montevideo: Instituto Nacional de Pesca, Ministério de Ganadería, Agricultura y Pesca - Programa de las Naciones Unidas para el Desarrollo. 1999. 79 p

[53] Shivji MS, Clarke S, Pank M, Natanson L, Kohler N, Stanhope M. Genetic Identification of Pelagic Shark Body Parts for Conservation and Trade Monitoring. Conservation Biology. 2002;**16**:1036-1047. DOI: 10.1046/j.1523-1739.2002.01188.x

[54] Pequeño G, Lamilla J. Las pesquerías de condrictios en Chile: primer análisis. Revista Biología Pesquera. 1997;**26**:13-24x

[55] Heist EJ, Gold JR. Genetic Identification of Sharks in the U.S. Atlantic large coastal Shark Fishery. Fishery Bulletin. 1999;**97**:53-61

[56] Gemaque R, Monteiro ILP, Gomes F, Sodré D, Sampaio I, Sales JBL, Rodrigues-Filho LFS. Why implement measures to conserve the diversity of Elasmobranchs? The case of the northern coast of Brazil. Revista da Biologia. 2017;**17**(2):1-7. DOI: 10.7594/revbio.17.02.01

[57] Lessa R, Santana FM, Rincón G, Gadig OBF, El-Deir ACA. Biodiversidade de Elasmobrânquios no Brasil. In: Relatório e Ações Prioritárias para Conservação da Biodiversidade da Zona Costeira e Marinha. 1999. 154 p

[58] Kyle CJ, Wilson CC. Mitochondrial DNA identification of game and harvested freshwater fish species. Forensic Science International. 2007;**166**:68-76. https://doi.org/10.1016/j.forsciint.2006.03.025

[59] Sevilla RG, Diez A, Noren M, Mouchel O, Jerome M, Verrez- Bagnis V, Van Pelt H, Favre-Krey L, Krey G, Consortium TF, Bautista JM. Primers and polymerase chain reaction conditions for DNA barcoding teleost fish based on the mitochondrial cytochrome b and nuclear rhodopsin genes. Molecular Ecology Notes. 2007;**7**:730-734. 10.1111/j.1471-8286.2007.01863.x

[60] Mendonça FF, Hashimoto DT, Porto-Foresti F, Oliveira C, Gadig OBF, Foresti F. Identification of the shark species Rhizoprionodon lalandii and R. porosus (Elasmobranchii, Carcharhinidae) by multiplex PCR and PCR-RFLP techniques. Molecular Ecology Resources. 2009;**9**(3):771-773. 10.1111/j.1755-0998.2009.02524.x

[61] Nachtigall PG, Rodrigues-Filho LF, Sodré DC, Vallinoto M, Pinhal D. A multiplex PCR approach for the molecular identification and conservation of the Critically Endangered daggernose shark. Endangered Species Research. 2017;**32**:169-175. DOI: https://doi.org/10.3354/esr00798

[62] Pinhal D, Shivji MS, Vallinoto M, Chapman DD, Gadig OBF, Martins C. Cryptic hammerhead shark lineage occurrence in the Western South Atlantic revealed by DNA analysis. Marine Biology. 2012;**159**:829-836. https://doi.org/10.1007/s00227-011-1858-5

[63] Rodrigues-Filho LFS, Rocha TC, Rêgo PS, Schneider H, Sampaio I, Vallinoto M. Identification and phylogenetic inferences on stocks of sharks affected by the fishing industry off the Northern coast of Brazil. Genetics and Molecular Biology. 2009; **32**(2):405-413. http://dx.doi.org/10.1590/S1415-47572009005000039

[64] Schmidt BF, Amorim AF, Hilsdorf AWS. PCR–RFLP analysis to identify four ray species of the genus Dasyatis (Elasmobranchii, Dasyatidae) fished along the southeastern and southern coast of Brazil. Fisheries Research. 20015;**167**:71-74. https://doi.org/10.1016/j.fishres.2014.12.025

[65] Tavares W, Rodrigues-Filho LFS, Sodré D, Souza RFC, Schneider H, Sampaio I, Vallinoto M. Multiple substitutions and reduced genetic variability in sharks. Biochemical Systematics and Ecology. 2013;**49**:21-29. https://doi.org/10.1016/j.bse.2013.02.004

Chapter 2

A Tale on the Demersal and Bottom Dwelling *Chondrichthyes* in the South of Sicily through 20 Years of Scientific Survey

Michele Luca Geraci, Sergio Ragonese,
Giacomo Norrito, Danilo Scannella,
Fabio Falsone and Sergio Vitale

Additional information is available at the end of the chapter

http://dx.doi.org/10.5772/intechopen.69333

Abstract

In the present work, an overview of the demersal (sharks-chimaera) and bottom dwelling (batoids) of experimental survey international bottom trawl survey in the mediterranean (MEDITS) data, from 1994 to 2013, is provided. The analysed data refer to a wide area located off the southern coast of Sicily, namely south of Sicily (according to the general fisheries commission for the mediterranean (GFCM) classification, Geographical Sub-Area 16). A checklist of the recorded *Chondrichthyes* was integrated by density index, D.I. (N/Km^2) and average individual weight (as the ratio between biomass index, D.I. (N/Km^2) and D.I.). Results suggest that most of the *Chondrichthyes* in South of Sicily are in a steady state, although in the last few years, they seemed to recover. The spatial distribution of sharks-chimaera in the geographical sub-area (GSA) 16 is mainly concentrated in the southern and north-western zones. Nevertheless, possible management actions to promote the recovering of these very important ecological and threatened species are discussed.

Keywords: sharks, chimaera, batoids, checklist, abundance, South of Sicily, Mediterranean Sea

1. Introduction

Marine cartilaginous species present unique challenges for conservation assessment (in Refs. [1, 2]). They are considered the most vulnerable species to fishing activity. Groundfish sharks and chimaera, together with bottom dwelling batoids, share several biological traits, for example: high position in the trophic food webs, slow growth, delayed sexual maturity, low fertility and long life spans. Generally, they form small local stock (the so-called stock-let) with limited or low connectivity to each other (in Refs. [3–5]). The assumption that marine fish are not vulnerable to extinction because they live in open seas where their movements are unlimited is unfounded. Fisheries have caused severe declines in many species, and although there are still no documented cases of complete extinction, there is considerable debate as to whether marine species could become extinct (in Refs. [2, 6, 7]).

In many areas of the world, a generalised decline of cartilaginous fish species standing stocks is generally recognized mainly because of fishing effort increase; the apparent paradox of a corresponding increase in landings of some stock likely reflects the reduction of the discarded catch due to the general crisis of the sector. Overfishing, habitat degradation and slow recovery rates are potential factors that lead to such dramatic declines, especially in areas such as the Mediterranean Sea where fishing (both legal and illegal) has long been a way of life and continues to be intense. As a matter of fact, *Chondrichthyes* in the Mediterranean Sea have always been considered as low-economic level bycatches and even a nuisance in the past and hence massively discarded, which is currently the case of *Galeus melastomus* in the South of Sicily (in Ref. [8]). That notwithstanding, they continue to decline as an indirect effect of fisheries aimed at more valuable species as generally recognized (in Refs. [2, 9–11].

In this context, current Mediterranean elasmobranchs are represented by ~85 shark and batoid species (in Ref. [12]) more or less in an over-exploited condition (i.e. standing stock very reduced); in particular, 31 species (40%) are regionally classified as threatened categories (critically endangered, endangered or vulnerable) (in Ref. [13]). The Mediterranean catches are multi-species with a rich marine community, including selachians, historically exploited by different fisheries (in Ref. [14]). In the wide area between the Southern Sicily and the Northern Coasts of Africa, *Chondrichthyes* have always represented a common bycatch of the bottom trawl fleets since the 70s. As observed worldwide, also in the above-reported area, bottom trawls are increasing in size, power and efficiency, as well as in their fishing activities, mainly towards the offshore grounds (in Ref. [4]). Due to the bycatch nature and high discard rate, long-term sources of information to assess *Chondrichthyes* gross catch are very limited in this region. In this context, fishing activity has determined a severe impact on *Chondrichthyes* communities for three main reasons. First of all, the aim of the fishermen was to increase the abundance of highly-prized shrimps through a regular harvest of low valuable demersal shark, a concept recognizable also in the old scientific literature, considering the sharks as predators of red shrimps and human's competitor (in Refs. [15–17]). However, successive studies have shown that sharks do not feed on red shrimps (in Refs. [3, 18, 19]).

Secondly, in the past poor data were recorded on these *taxa* since the scattered retention of large specimens as well as few categories were documented in the official Italian statistics (in Refs. [20, 21]). Nowadays, a dedicated regulation and data collection is in place (in Ref. [22]).

Thirdly, the scarce selectivity of the commercial trawling cod-end (diamond, 20–30 mm side stretched; in Ref. [19]) together with the 5–6 hours for haul, has determined a huge catch of *Chondrichthyes*, mainly rejected at the sea. The fishing activity above-reported took place in the past (in Ref. [23]) and, although the Reg. (EC) n. 1967/2006 imposed an increased mesh size (square, 40 or 50 mm diamond) in the cod-ends, almost all the shark specimens continue to be retained by the cod-end (in Ref. [24]). Regarding the GFCM geographical Sub-Area no 16, named South of Sicily, a standardized scientific data base was obtained after the implementation of MEDITS international program, launched in 1994, and based on a high vertical opening (ca 2–3 m) trawl net.

To improve the knowledge on *Chondrichthyes* occurring in the South of Sicily, a time series of data collected during the experimental MEDITS survey were analysed.

2. Overview of the *Chondrichthyes* knowledges

The used data were gathered during MEDITS survey program and specifically referring to the South of Sicily (geographical sub-area (GSA) 16 according to GFCM classification). This area extends for about 34,000 km^2 and is characterized by the entry of the modified Atlantic Water (AW), which flows towards east in proximity of the surface (up to around 200 m), and from the spillage of warmer and salty water (200–500 m), the levantine intermediate water (LIW), which flows towards west, along the Sicilian slope (**Figure 1**).

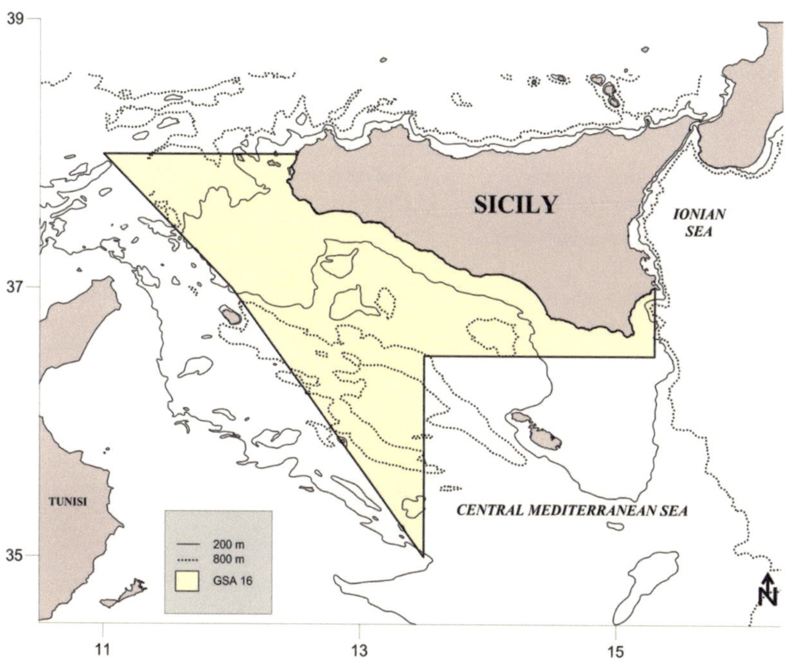

Figure 1. The geographical zone considered in the present paper is South of Sicily (GSA 16) with an overall extension of about 34,000 km^2. The solid and dotted lines denote the 200 and 800 m depth levels.

In the investigated area, bottom trawling is forbidden (but such a measure is rarely enforced within 50 m of depth) from the Sicilian coasts and in some specific grounds (such as the Egadi Islands marine protected area, MPA). The data referring to the MEDITS Survey from 1994 to 2013 were carried out with a commercial stern trawler harboured in Mazara del Vallo, the Sant'Anna (32.2 m length overall; powered with a 736 kW engine). The sampling stations have been distributed applying a stratified sampling scheme with random drawing inside the following bathy-metric limits: 10–50 m (a *stratum*), 51–100 m (b), 101–200 m (c), 201–500 m (d), 501–800 m (e). 30 and 60 minutes day light hauls were performed on shelf (10–200 m) and slope (201–800 m) grounds respectively. The mean D.I. performed through number of hauls per year and the ratio between B.I./D.I. were computed for each species collected at least once in the MEDITS surveys and consequently pooled together.

In the present study, a checklist of *Chondrichthyes* recorded in the GSA 16 during the MEDITS survey was also produced and presented in phylogenetic order and within this in alphabetic order. Furthermore, an overall D.I. maps for sharks-chimaera and batoids as well as their spatial occurrences are displayed. The spatial analysis was performed using geostatistical methods (ordinary kriging) across the timeframe from 1994 to 2013.

In the South of Sicily, overall 37 species were recognized as captured at least once: 16 demersal sharks-chimaera and 21 batoids. In particular, four sharks-chimaera orders (*Chimaeriformes, Hexanchiformes, Squaliformes* and *Carcharhiniformes*; **Table 1**) and three batoids orders (*Myliobatiformes, Rajiformes* and *Torpediniformes*; **Table 2**) were found. Synthetic comments concerning the 16 sharks-chimaera and 21 batoids *taxa* are reported hereafter in phylogenetic order.

Class: Chondrichthyes			
Order *Carcharhiniformes*			
Family *Scyliorhinidae*	*Galeus melastomus* Rafinesque, 1810a	*Scyliorhinus canicula* Linnaeus	*Scyliorhinus stellaris* Linnaeus, 1758LC
Family *Triakidae*	*Mustelus asterias* Cloquet, 1821	*Mustelus mustelus* Linnaeus, 1758	*Mustelus punctulatus* Risso 1827
Order *Chimaeriformes*			
Family *Chimaeridae*	*Chimaera monstrosa* Linnaeus, 1758		
Order *Hexanchiformes*			
Family *Hexanchidae*	*Heptranchias perlo* Bonnaterre, 1788	*Hexanchus griseus* Bonnaterre, 1788	
Order *Squaliformes*			
Family *Centrophoridae*	*Centrophorus granulosus* Bloch and Schneider, 1801	*Centrophorus uyato*	
Family *Dalatiidae*	*Dalatias licha* Bonnaterre, 1788		
Family *Etmopteridae*	*Etmopterus spinax* Linnaeus, 1758		
Family *Oxynotidae*	*Oxynotus centrina* Linnaeus, 1758		
Family *Squalidae*	*Squalus acanthias* Linnaeus, 1758	*Squalus blainville* Risso, 1827	

Table 1. Checklist of the sixteen *taxa* (sharks-chimera) caught during the MEDITS survey from 1994 to 2013 in the South of Sicily (GSA 16).

2.1. Blackmouth catshark—*G. melastomus* Rafinesque, 1810

This small-sized shark lives in deep waters from 150 to more than 2000 m (in Ref. [25]), even though it can be occasionally (especially juveniles) found over the inner shelf (50–60 m) (in Refs. [25, 26]). In the Mediterranean, it reaches a maximum size of 63 cm total lenght (TL) (in Refs. [27]). *G. melastomus* feeds mainly on shelf-living species, natantian and reptantian crustaceans together with teleosts (in Ref. [28]). Analysis of the stomach contents of individuals caught in the northern Tyrrhenian Sea also found galley leftovers (in Ref. [28]). Blackmouth catshark is generally considered as having an awful taste and Sicilian fishers discard it immediately (in Refs. [8, 17, 29]) with a very low possibility of surviving.

2.2. Small-spotted catshark—*Scyliorhinus canicula* Linnaeus, 1758

A bottom dweller shark measuring up to 90–100 cm of TL (in Refs. [30, 31]) lives in gravel, sandy and muddy bottoms down to 800–1000 m, but preferably within the 400–500 m depth range (in Refs. [32, 33]). The Small-spotted catshark is an opportunistic predator on a wide range of macrobenthic fauna (generalist feeder). In particular, natantian and reptantian crustaceans together with teleosts were the most important preys (in Ref. [28]). *S. canicula* is taken in commercial fisheries across its range but only larger individuals are usually retained for human consumption; the juveniles discarded seem to have some possibility to survive.

2.3. Nursehound—*Scyliorhinus stellaris* Linnaeus, 1758

Its habits are similar to the Small-spotted catshark, but it prefers the rocky zones from 20 to 100 m (in Refs. [30, 31]), although it is even able to go down to 800 m (in Ref. [30]). In the Mediterranean, its maximum size is 150 cm TL (in Ref. [34]). The Nursehound feeds on benthic prey, mainly on crustaceans, molluscs, some bony fishes and on its congeneric small spotted catshark, *S. canicula* (in Ref. [31]). It is not appreciated by Sicilian consumers and so often discarded (in Ref. [29]). Nowadays, it has almost disappeared in many Sicilian fishing grounds and remains common only in Tunisian (in Ref. [26]) and Maltese waters (in Ref. [35]).

2.4. Starry smooth-hound—*Mustelus asterias* Cloquet, 1821

A slender shark measuring up to 140 cm TL (in Ref. [31]), though occurring below, from a few metres to about 100 m (in Ref. [32]), can be caught below 300 m (in Ref. [30]) and sometimes deeper at 500 m (in Ref. [34]). Starry Smooth-hound feeds predominantly on crustaceans, including squat lobsters and crabs, and especially swimming crabs. Predation on other taxa is low (in Ref. [36]). In the Mediterranean Sea, trawling and artisanal fishing have largely exploited the *Mustelus* species that were regularly commercialized for human consumption (in Ref. [37]).

2.5. Smooth-hound— *Mustelus mustelus* Linnaeus, 1758

This species measures up to 160 cm TL (in Ref. [34]) and lives at 800 m depth (in Ref. [32]), but shows a preference for shallow sandy-muddy bottoms, especially at 5–50 m depth (in Refs. [30, 31]. Gracan et al. (in Ref. [38]) declared that *M. mustelus* feeds mainly on crabs. Decapod malacostracans were the dominant prey group, with *Liocarcinus corrugatus*,

Liocarcinus depurator and *Pilumnus sp.* as the most frequent prey; they were followed by ray-finned fish, mostly *Engraulis encrasicolus* and cephalopods, consisting mainly of *Sepia elegans*. In the Mediterranean Sea, smooth-hound is captured with demersal trawls, trammel nets, gillnets and longlines. It is a commercial species, mostly taken as bycatch and marketed (in Ref. [4]).

2.6. Blackspotted smooth-hound—*Mustelus punctulatus* Risso, 1827

As the similar species of *M. mustelus* (with which it is often confused (in Ref. [39]), this shark measures up to 190 cm TL and is reported in the whole Mediterranean (in Ref. [30]). *M. punctulatus* feeds mainly on crustaceans, teleosts and molluscs although its diet changes during the life cycle (in Ref. [39]). Like the other, *Mustelus* species is captured, as bycatch, with demersal trawls, trammel nets, gillnets and longlines and often landed (in Ref. [36]).

2.7. Rabbitfish—*Chimaera monstrosa* Linnaeus, 1758

This Atlanto-Mediterranean deep-water animal prefers cold waters and occurs in all the Mediterranean, except the North Adriatic (in Ref. [40]). Its depth limits range from the outer shelf down to ca. 1600 m (in Ref. [40]). Maximum length is 150 cm TL and maximum age is 26 and 30 years for females and males, respectively (in Ref. [41]). *C. monstrosa* feeds on benthic organisms and it has a very close relationship with the seabed in its feeding habits. The diet is composed mainly by crabs, particularly the angular crab *Goneplax rhomboides* which represents nearly half of the diet by volume, ophiuroids, echinoids, crinoids, amphipods, polychaetes, pagurids, cnidarians and other medium-sized benthic prey (in Refs. [42, 43]). In the Mediterranean, *C. monstrosa* is usually caught by offshore trawlers mostly between 500 and 800 m (in Refs. [32, 44]) and immediately discarded (in Ref. [8]).

2.8. Sharpnose sevengill shark—*Heptranchias perlo* Bonnaterre, 1788

This shark is easily recognizable for the presence of seven gill slits and occurs in the whole Mediterranean showing a wide depth-distribution from 0 to 50 m down to 800 to 1000 m (in Refs. [8, 31]). The newborn is ca. 30 cm TL and during adulthood reaches 90–100 cm TL, attains a maximum size of up to 140 cm TL (in Ref. [31]) and feeds on small sharks and rays, small bony fish, shrimps, crabs, lobsters, squid and cuttlefish (in Ref. [45]). It is taken by a wide variety of demersal fisheries and sold at the supermarket.

2.9. Bluntnose sixgill shark—*Hexanchus griseus* Bonnaterre, 1788

Bluntnose sixgill shark lives up to 2500 m depth (in Ref. [46]) although during the night it was noticed at a depth around 30–40 m in the Straits of Messina (in Ref. [47]). The maximum TL recorded was 600 cm (in Ref. [46]). This shark is taken as bycatch in handlines, longlines, gillnets, traps, trammel nets, and both mid-water and bottom trawls. There are some small-scale fisheries for this species in the Mediterranean (in Ref. [46]). In the mid 80s, large sized animals were commonly found at the fish market in Mazara (Sicily) and sold as slices of 'Palumbo' (the Sicilian name for *Mustelus* spp.). Nowadays, it's not sold and discarded at sea (in Ref. [4]).

http://dx.doi.org/10.5772/intechopen.69333

2.10. Gulper shark—*Centrophorus granulosus* Bloch and Schneider, 1801

A common deep-water species (often confused with the congener *Centrophorus uyato*, see below) which grows up to 120 cm TL (in Ref. [30]) and lives in a depth range from 50 to 1400 m. The diet of the Gulper shark is poorly understood but it is thought to prey on hake, lanternfish, squid and epigonids, as well as a variety of other benthic and mesopelagic bony fish and invertebrates (in Ref. [48]). Marketed smoked and dried salted for human consumption; also processed into fishmeal and a source of liver oil for squalene (in Ref. [31)].

2.11. Little Gulper shark—*Centrophorus uyato* Linnaeus, 1758

Demersal on the continental shelf and upper-middle continental slope at depths of 50 to 1,400 m, This invalid *taxon* was distinguished from the similar species *C. granulosus* mainly according to the shape of the superior teeth and features of the dermal denticles on the sides of the body (in Refs. [23, 31]. Maximum recorded TL is 110 cm (in Ref. [49]). The diet consists of bony fishes and cephalopods (in Ref. [50]) but also includes crustaceans (in Ref. [51]). However, Sicilian fishers do distinguish between the two 'forms', which are often landed and commercialized (in Ref. [4]).

2.12. Kitefin shark—*Dalatias licha* Bonnaterre, 1788

A benthic to mesopelagic deep-water shark occurring at depths between 90 and 1400 m (in Ref. [52]), which grows up to 180 cm TL (in Ref. [30]); however, 120 cm is a more common length (in Ref. [34]). Navarro et al. (in Ref. [53]) revealed a preference for small sharks; however, finfish, crustaceans and cephalopods were also found. The species occurs within the range of fisheries in many areas of its range, where it is taken as bycatch. It is sometimes sold at the fish markets, but normally it is discarded (especially by Sicilian red shrimp trawlers) (in Ref. [8]).

2.13. Velvet belly—*Etmopterus spinax* Linnaeus, 1758

This Atlanto-Mediterranean deep-water shark has a benthic life on the shelf and bathyal zones, from 70 to about 2500 m (in Ref. [54]) but mostly below 200 m depth (in Ref. [32]). *E. spinax* fed mainly on cephalopods (in Ref. [28]). The diet of *E. spinax* was composed primarily of mesopelagic fish, with decapod crustaceans and cephalopods of secondary importance (in Ref. [55]). A non-commercial species, all specimens captured as bycatch by commercial fishing vessels are discarded thus limiting the data available (in Ref. [56]).

2.14. Angular roughshark—*Oxynotus centrina* Linnaeus, 1758

Angular roughshark lives in a bathymetric range from 60 to 660 m (in Ref. [31]); however, it can reach a depth up to 800 m (in Ref. [44]). The maximum TL recorded was 150 cm (in Ref. [30]). The reported diet of this species is mainly characterized by small crustaceans (in Ref. [57]), polychaetes (in Ref. [31]) and teleosts (in Ref. [58]). Guallart et al. (in Ref. [59]) considered *O. centrina* to be a suction feeder specializing in worm-like prey. It is caught by trawling or accidentally bottom longlining, from a few metres to deeper waters. It is immediately discarded to the sea by fishers from Mazara because they think it will bring bad luck (in Ref. [4]).

2.15. Piked dogfish—*Squalus acanthias* Linnaeus, 1758

Piked dogfish is a small bottom-dwelling shark with a maximum recorded size of 160 cm TL and maximum weight of 91 kg (in Ref. [31]) with a maximum depth of 800 m (in Ref. [60]). It can be considered an opportunistic feeder. Their natural diet composed mainly of teleost fishes, followed by crustaceans, nematodes and actinarians (= sea anemones) (in Ref. [61]). It's taken as bycatch in demersal fisheries and sold at supermarket (in Ref. [4]).

2.16. Longnose spurdog—*Squalus blainville* Risso, 1827

It is a small shark measuring up to 110 cm TL and occurring at 700 m depth (in Ref. [60]). In the stomach, contents of *S. blaiville* crustaceans and teleosts were the dominant prey items, and molluscs, polychaetes, echinoderms and sipunculids were found in lower abundance (in Ref. [62]). *S. blainvellei* is of limited fisheries importance compared to *S. acanthias*, but may also have been impacted by fishing pressure in this area (in Ref. [63]). It is very common and sold at the supermarket (in Ref. [4]).

Class: Chondrichthyes			
Order *Myliobatiformes*			
Family *Dasiatidae*	*Dasyatis pastinaca* Linnaeus, 1758	*Pteroplatytrygon violacea* Bonaparte, 1832	
Family *Myliobatidae*	*Myliobatis aquila* Linnaeus, 1758	*Pteromylaeus bovinus* Geoffroy St. Hilaire, 1817	
Order *Rajiformes*			
Family *Rajidae*	*Dipturus batis* Linneo, 1758	*Dipturus oxyrinchus* Linneo, 1758	*Leucoraja circularis* Couch, 1838
	Leucoraja fullonica Linneo, 1758	*Leucoraja melitensis* Clark, 1926	*Leucoraja naevus* Muller & Henle, 1841
	Raja asterias Delaroche, 1809	*Raja brachyura* Lafont, 1873	*Raja clavata* Linneo, 1758
	Raja miraletus Linneo, 1758	*Raja montagui* Fowler, 1910	*Raja polystigma* Regan, 1923
	Raja radula Delaroche, 1809	*Rostroraja alba* Lacépède, 1803	
Order *Torpediniformes*			
Family *Torpedinidae*	*Torpedo marmorata* Risso, 1810	*Torpedo nobiliana* Bonaparte, 1835	*Torpedo torpedo* Linnaeus, 1758

Table 2. Checklist of the twenty-one taxa (batoids) caught during the MEDITS survey from 1994 to 2013 in the South of Sicily (GSA 16).

2.17. Common stingray—*Dasyatis pastinaca* Linnaeus, 1758

It occurs from the shore to about 200 m depth, but is more commonly found in shallow waters <50 m (in Refs. [12, 64, 65]). It feeds on a wide variety of bottom-dwelling organisms. In a study by Ismen [66], crustaceans represented more than 99% of the diet when pooling all size classes, but teleost fish were of increasing importance in the diet of larger stingrays. The common stingray has been reported to reach a disc width (DW) of 1.4 m and a TL of 2.5 m, though a DW of 45 cm is more typical. Common stingrays are caught incidentally by commercial fisheries across many parts of its range, using bottom trawls, gillnets, bottom longlines, beach seines, and trammel nets (in Ref. [30]).

It is discarded after fishermen cut off the dangerous tails (in Ref. [67]), which have caused at least one fatality among fishers from Mazara.

2.18. Blue stingray—*Pteroplatytrygon violacea* Bonaparte, 1832

It occurs from over the edge of continental and insular shelves into the open water but has been reported at 238 m depth. In the Adriatic water, the diet consisted of two main taxonomic groups such as teleost fish and cephalopods, but few specimens of crustaceans were also recorded (in Ref. [68]). This species is captured by pelagic longline fisheries operating in the Mediterranean Sea (in Ref. [68]). *P. violacea* is almost certainly the most discarded elasmobranchs. In Italian seas, pelagic stingrays are the most commonly caught elasmobranch species in the Albacore long line fisheries and the second most common elasmobranch catch in swordfish long line fisheries (in Ref. [69]).

2.19. Common eagle ray—*Myliobatis aquila* Linnaeus, 1758

The common eagle ray in the Mediterranean is reported on sandy and muddy substrates, from shallow water to 200 m depth, although it was reported at the depth of 537 m off southern African coast (in Ref. [70]). It is a relatively small ray, attaining a maximum size of 80 cm DW (in Ref. [71]). It feeds on invertebrates such as crabs, mole crabs and bivalves, and on small bony fishes. The wings are said to be good eating and along the African coast are regularly used for human consumption. The species represent a regular bycatch in mixed species fisheries (in Ref. [70]).

2.20. Bull ray—*Pteromylaeus bovinus* Geoffroy St. Hilaire, 1817

It has a moderate depth range from costal water to about 30 m, occasionally in oceanic water up to100 m of depth. In Eastern Mediterranean Sea, Dulcic et al. (in Ref. [72]) found a max TL of 2940 cm and 220 cm DW for female. Regarding feeding aspects, it is known that they prey on bottom-living invertebrates such as crustaceans (crabs, prawn) and mollusks (squids, bivalve) (in Ref. [73]). Bull rays are very rare and not commonly caught by fisherman; they are mainly discarded at sea (in Ref. [74]).

2.21. Gray skate—*Dipturus batis* Linneo, 1758

Gray skate is found from shallow coastal waters to depths of 600 m, but most commonly found at 200 m depth. Maximum-recorded TL is 250 cm (in Ref. [30]). *D. batis* preys mostly on crustaceans and teleost fish, although Steven (in Ref. [75]) reported several species of elasmobranch, including other species of rajid, in the stomach contents. It is caught as bycatch of multi-species trawl fisheries, which cover much of its shelf and slope habitat. It is mainly landed in northern Europe where it is fished by trawlers and longliners (in Ref. [30]).

2.22. Longnosed skate—*Dipturus oxyrinchus* Linneo, 1758

Longnosed skate is found in water from 90 to 900 m, commonly around 200 m (in Ref. [76]). In the Mediterranean Sea, typical TL varies between 60 and 100 cm but it can reach a maximum TL of 150 cm (in Ref. [77]). The diet comprised crustaceans and molluscs. Early life stages were characterized by a benthic diet, which changed to benthopelagic during growth (in Ref. [78]). *D. oxyrinchus* is captured as part of the bycatch of multispecies trawl fisheries.

2.23. Sandy ray—*Leucoraja circularis* Couch, 1838

Demersal on sandy and muddy bottoms from the outer shelf and upper slope to 275 m depth, commonly found at 100 m depth (in Ref. [30]), maximum recorded size is 120 cm TL, but most individuals caught are between 70 and 80 cm TL (in Ref. [30, 65]). Its diet is poorly understood but it is most likely that it feeds on various bottom dwelling invertebrates, particularly crustaceans, and small teleost fish (in Ref. [79]). Species of local fishery importance are caught by bottom trawl fisheries.

2.24. Shagreen ray—*Leucoraja fullonica* Linneo, 1758

Demersal on rough ground on outer shelf and upper slope in about 30 to 550 m depth, maximum length is 120 cm; however, most specimens usually are 70 to 80 cm TL (in Ref. [65]). It feeds on a variety of bottom dwelling species but most probably prefers fish and crustaceans (in Ref. [80]). In the Mediterranean Sea, it is caught as bycatch by both bottom trawl and longline fisheries (in Ref. [30]).

2.25. Maltese ray—*Leucoraja melitensis* Clark, 1926

Maltese ray deep range is from 60 to 600 m. This small ray grows up to 50 cm TL (in Ref. [30]). It feeds on crustaceans mainly amphipods (in Ref. [85]). While this skate is not known to be targeted by commercial fisheries, it is taken as bycatch in bottom trawl, gillnet, and bottom longline fisheries and often discarded (in Ref. [81]).

2.26. Cuckoo ray—*Leucoraja naevus* Muller & Henle, 1841

Demersal on sandy and course bottoms on the shelf from 30 to 200 m depth, it is found on the continental shelf and slope at depths of 20–500 m (in Ref. [30]), but it is most common between

50 and 200 m depth (in Ref. [65]). The maximum TL is 75 cm for females and 68 cm for males (in Ref. [65]). Juvenile Cuckoo Rays feeds mainly on small crustaceans while adults also feed on polychaetes and bony fish (in Ref. [82]). Bycatch of bottom trawl fisheries (in Ref. [30]).

2.27. Starry ray—*Raja asterias* Delaroche, 1809

Starry skates are found predominantly on the Italian and Corsican continental shelves between shallow waters and 150 m depth. Maximum size is estimated at 72 cm TL and 76 cm TL for males and females, respectively (in Ref. [83]). *R. asterias* is a predator of crustaceans (crabs and shrimps) and to a lower extent on teleosts, molluscs and polychaetes (in Ref. [84]) and is regularly caught as bycatch in the bottom trawl fisheries. As secondary target species is caught by beam trawl, juvenile specimens are frequently caught by trammel net in very shallow waters (2–15 m) and discarded alive (in Ref. [30]).

2.28. Blonde ray—*Raja brachyura* Lafont, 1873

Demersal on sandy grounds from inshore to upper slope exceptionally as deep as 900 m, it reaches a maximum size of ~120 cm (TL) and commonly reaches 40–100 cm TL (in Ref. [85]). Fish were a major prey item for all sizes of *R. brachyura*. Excluding bony fish, polychaetes were the most common prey followed by shrimps and brachyuran crabs and cephalopods (in Ref. [82]). It is taken as bycatch in mixed demersal fisheries using trawl, gill nets and longlines elsewhere in its range (in Ref. [86]).

2.29. Thornback ray—*Raja clavata* Linneo, 1758

A relatively common skate from close in-shore shallow waters to the outer continental shelf and upper slope from 10 to 300 m depth (in Ref. [77]), maximum-recorded TL is 110 cm (in Ref. [87]). Thornback ray feeds mainly on teleosts, crustaceans and cephalopods, whereas gastropods and polychaetes are occasionally consumed (in Ref. [88]). In the Mediterranean Sea, the Thornback skate is frequently caught as bycatch in trawl fisheries targeting the Rose Shrimp (*Parapenaeus longirostris*) and the European Hake (*Merluccius merluccius*). There are localized and targeted fisheries for this species in the Mediterranean Sea (in Ref. [89]). Recently, Bottari et al. (in Ref. [93]) have supported the stock in GSA 16 as a different Unit stock from the surrounding GSAs.

2.30. Brown ray—*Raja miraletus* Linneo, 1758

Demersal on soft bottom from shallow shelf to about 530 m depth, mainly at 50 to 150 m (in Ref. [30]), *R. miraletus* is a small ray that can reach maximum TL of 60 cm (in Refs. [30, 90]). In Brown ray diet, crustacean represented the main prey in all size groups. Amphipoda Gammaridea was the predominant prey for small individuals and was replaced in medium and large specimens by decapods (in Ref. [91]). It is caught as bycatch in bottom trawl, trammel net and long-line fisheries. It is landed and commonly sold in the market (in Ref. [92]).

2.31. Spotted ray—*Raja montagui* Fowler, 1910

Demersal on soft substrate on shelf at 30–150 m depth, rarely as deep as 530 m, most individuals attain a TL of 40–60 cm (in Ref. [34]) and the maximum-recorded TL is 80 cm (in Ref. [94]). For *R. montagui,* most important preys were various crustaceans and polychaetes, while large females predated primarily on fish (in Ref. [82]). *R. montagui* is captured in Mediterranean trawl fisheries as bycatch (in Ref. [95]).

2.32. Speckled ray—*Raja polystigma* Regan, 1923

R. ploystigma is caught predominantly on the shelf living on soft bottoms from 100 to 400 m (in Refs. [30, 95]). Its maximum size is about 50–60 cm TL and it feeds mainly on crustaceans and bony fishes depending on the sex, size and also partly on the season (in Ref. [28]). The species is caught as bycatch in demersal trawl fisheries but is also fished with gillnets, longlines and handlines in artisanal fisheries (in Ref. [34]).

2.33. Rough ray—*Raja radula* Delaroche, 1809

Rough ray occurs in coastal water up to 350 m depth (in Ref. [30]). Maximum size is about 50–60 cm TL (in Refs. [30, 76]). The diet of juvenile specimens of *R. radula* consisted mainly of crustacean decapods followed by algae, polychaetes and molluscs (in Ref. [96]). *R. radula* is frequently caught as bycatch in demersal trawl, gillnet, trammel net and bottom longline fisheries (in Refs. [30, 96]).

2.34. White skate—*Rostroraja alba* Lacépède, 1803

White skate is a demersal species found on the continental shelf and upper slope from shallow water to 400 m, exceptionally to 500 m depth on sand and loose rocky substrate (in Refs. [30, 85]). Recorded maximum TL is 200 cm, though common between 60 and 150 cm of TL (in Refs. [34, 97]). Regarding diet aspect, it is known to prey mainly on fish and to lesser extent on crustaceans (in Ref. [98]). It is taken mainly as bycatch of bottom trawl fisheries (in Ref. [97]).

2.35. Marbled electric ray—*Torpedo marmorata* Risso, 1810

Marbled electric ray lives in inner shelves on soft and stony bottom to about 40 m depth, rarely deeper to about 100 m (in Refs. [30, 65]). *T. marmorata* commonly grows to 40 cm TL (in Ref. [99]) but it may grow up to 100 cm (in Ref. [30]). Marbled electric ray is an active feeder, consuming mostly fish and to a lesser extent cephalopods (in Ref. [100]). Bycatch of bottom trawls demersal fisheries in coastal grounds (in Ref. [30]).

2.36. Electric ray—*Torpedo nobiliana* Bonaparte, 1835

T. nobiliana occurs from the surface to depths of ~800 m. Juveniles are mainly benthic occurring on soft substrates and coral reef habitats, from 10 to 50 m depth (in Ref. [32]). It is the biggest of Mediterranean Electric Ray and may reach the notable size of 180 cm TL (in Ref. [65]). The diet is predominantly fish, sometimes quite large. Usually discarded at the sea, it is sometimes a bycatch in bottom trawl and artisanal demersal fisheries in coastal grounds (in Ref. [65]).

2.37. Common torpedo—*Torpedo torpedo* Linnaeus, 1758

Mainly a benthic species found in near shore habitats and on soft bottoms, but also to about 70 m depth and occasionally deeper. *T. torpedo* TL usually ranges from 44 to 47 cm TL (in Ref. [101]) but it can reach about 60 cm TL (in Ref. [30]). Common torpedo is an active feeder, consuming mostly fish and to a lesser extent crustaceans (in Refs. [102, 103]). Bycatch in bottom trawls coastal fisheries (in Ref. [30]).

3. State of the art of *Chondrichthyes* in South of Sicily

Among the seven orders of *Chondrichthyes* (*Chimaeriformes, Hexanchiformes, Squaliformes, Carcharhiniformes, Myliobatiformes, Rajiformes* and *Torpediniformes*) detected in the GSA 16, 37 species are under discussion in order to better define the right systematic position or taxonomic issue. The doubts are mainly due to the similar morphological and morphometric characters. For example, in the Gulf of Gabès (southern Tunisia), the species status of longnose spurdog has been questioned, and in fact the meristic data along with genetic analysis support the assignation of longnose spurdog to shortnose spurdog (*Squalus megalops,* Macleay) (in Refs. [12, 104]). Another famous case of systematic confusion relates to the already cited Gulper sharks *C. granulosus* and *C. uyato*; both were listed as valid species for a region but it has to be noted that the whole genus needs revision worldwide (in Ref. [105]).

Regarding batoids, another potential taxonomic misidentification could be related to speckled ray and spotted ray as well as to marbled stingray (*Dasyatis marmorata,* Steindachner, 1882) and common stingray.

In conclusion, for many *Chondrichthyes*, there seems to be required an improvement on the taxonomic issue through genetic studies, with the aim to formally resolve the uncertainty identification.

In the investigated area, the analysis of the D.I. and B.I./D.I. temporal evolution from 1994 to 2013 highlights a slight recovery of sharks-chimaera (**Figure 2**) while it seems to be in steady state for batoids (**Figure 3**).

Observing the temporal evolution of the sharks-chimaera D.I. a stable trend is pointed out up to 2003, while a marked increment is underlined until 2008, while in the remaining five years, the D.I. seems fluctuate. Regarding batoids, the D.I. seems to fluctuate, although a clear increase is recorded between 2003 and 2010. The B.I./D.I. ratio seems more heterogeneous for batoids. This might be due to the different gear recruitment between the two investigated taxa as well as behaviour aspects (e.g. aggregation, swimming capability, feeding habits, etc.), relation with the bottom and life history traits. Detailed knowledge of elasmobranch habitat requirements is essential for biodiversity conservation and fisheries management, but this is often hampered by a poor understanding of their spatial ecology (in Ref. [5]). Indeed, the

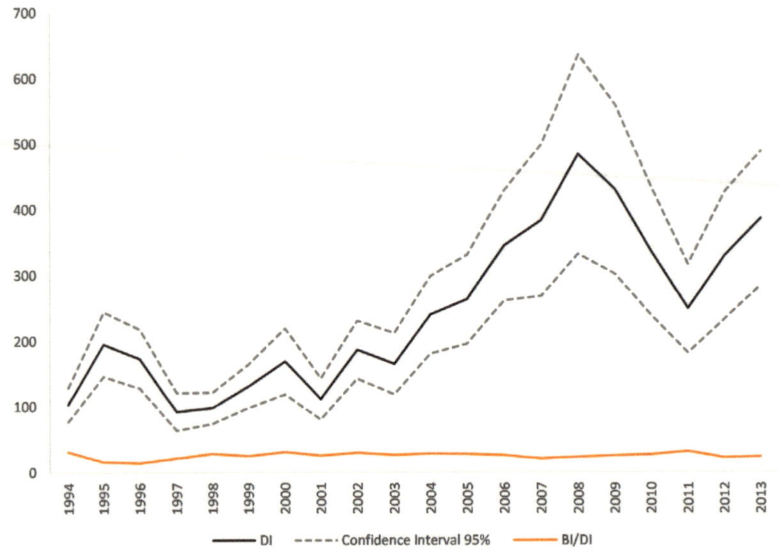

Figure 2. Annual (X ax) trend of density index (D.I. N/km^2; Y ax) and ratio between biomass index (B.I. Kg/km^2)/D.I. (Y ax), averaged across all the species, for the retained sharks and chimaera during the MEDITS survey in South of Sicily (GSA 16).

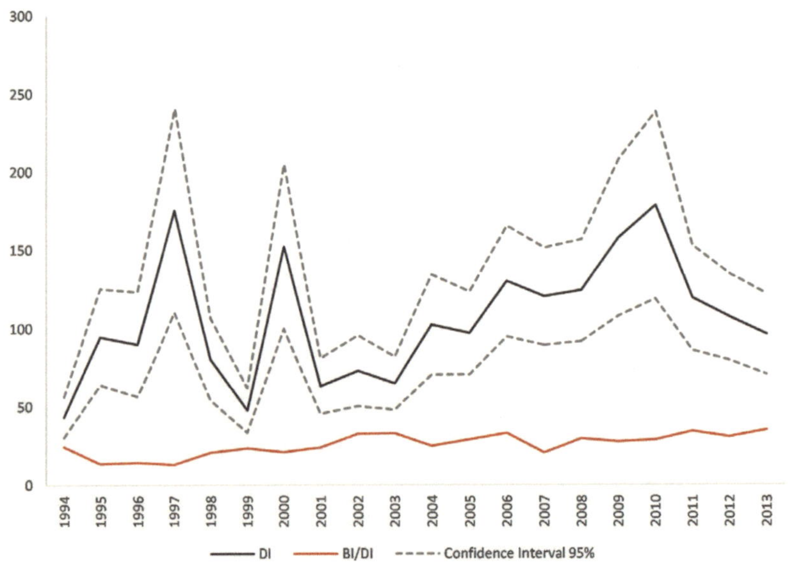

Figure 3. Annual (X ax) trend of density index (D.I. N/km^2; Y ax) and ratio between biomass index (B.I. Kg/km^2)/D.I. (Y ax), averaged across all the species, for the retained batoids during the MEDITS survey in the South of Sicily (GSA 16).

trends displayed above suggest that excluding the traditional considered rare species (such as *O. centrina, P. violacea, D. batis*, etc.), the response to fishing activities is not always the same. Valuable differences in resilience might be mainly related to the interaction between different factors (e.g. overall body shape, surviving capabilities after discarding, commercial value, etc.). In this context Ragonese et al. (in Ref. [81]) highlighted among sharks the most vulnerable and prone to decline and local extinction seems to be the neritic (such as *S. stellaris*), ovoviviparous and valuable/appreciated sharks (*Mustelus spp.* and *Squatina spp.*). *Mustelus spp*, which exhibit a big curiosity towards divers, and are also heavily speared by recreational fishers (Ragonese pers.obs.) Regarding batoids, Bradai et al. (in Ref. [12]) underline some neritic species that are almost disappeared locally (*e.g. R. alba*) or highly depleted (*R. polystigma*), whereas few species are quite stable (e.g. *R. clavata; R. miraletus*) although in a depressed abundance (in Ref. [93]).

The spatial distribution of sharks-chimaera in the GSA 16 (**Figure 4**) is mainly concentrated in the southern and north-western zones. The D.I. distribution is characterized by several patches reaching values until 1600 N/km^2, although in the north-western a hotspot is recorded with values up to 2400 N/km^2. This pattern mainly reflects the abundance and distribution of the small catsharks *S. canicula* (which prefers the outer shelf)) and *G. melastomus* (which is more abundant next to the deep basins of GSA 16) (in Ref. [25]).

Compared to sharks-chimaera, the spatial distribution of batoids (**Figure 5**) is more circumscribed within two main zones, one biggest in the north-western and the other in the southern part of the GSA 16. A small patch is recognized along the Sicilian coast with D.I. values up to 1200 N/km^2. In the north-western zone, higher values of D.I. (1600 N/km^2) are recorded.

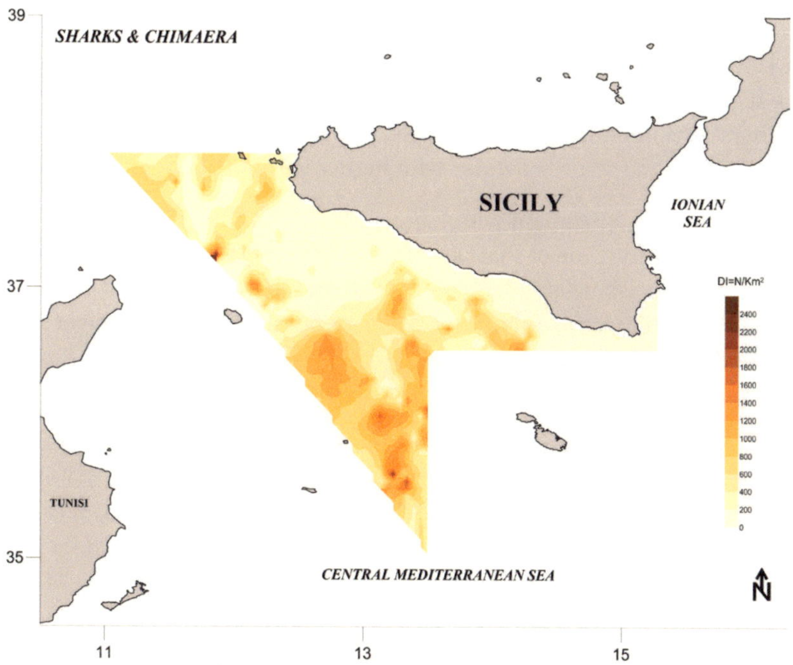

Figure 4. Spatial distribution of sharks-chimaera density index (D.I. N/km^2) in the South Sicily (GSA 16) from 1994 to 2013.

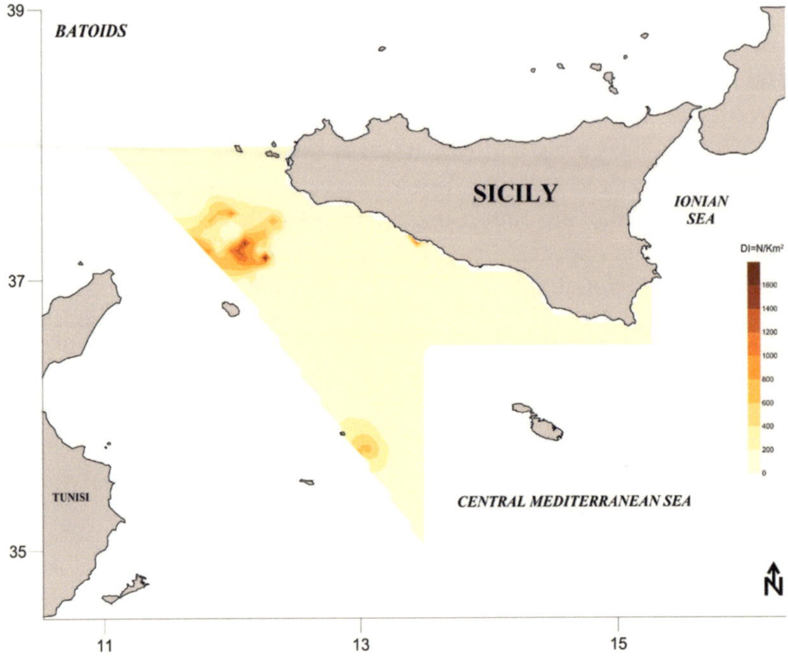

Figure 5. Spatial distribution of batoids density index (D.I., N/km^2) in the South Sicily (GSA 16) from 1994 to 2013.

Taking into account the published data on fishing effort (F) recorded by vessel monitoring system (VMS) in the years 2006–2010 (in Ref. [106]), an astonishing overlap is observed between the highest rate of fishing effort and the lowest of D.I. for both taxa. The above-mentioned considerations emphasize the importance to adopt an eco-sustainable fishery in the near future, at least in the recorded zones with higher values of D.I.

The present results are in agreement with Ragonese et al. (in Ref. [4]) who reported that in the same investigated area, the state of sharks and chimaera seems quite stable or even improving (for some species). In the same area Gancitano et al. (in Ref. [107]), carried out a similar study considering all the cartilaginous fishes together. Similarly, a slight increase of D.I. trend was displayed from the first 2000s until 2008 although a marked increase is showed by two years forward predications (2014–2015).

In the central Mediterranean Sea, Lauria et al. (in Ref. [5]) implemented habitat models considering only species with percentage of occurrence (always using MEDITS data) >5% and defined as 'Near Threatened', 'Vulnerable', 'Critically Endangered' or 'Data Deficient' in the international union for conservation of nature (IUCN) Red Lists. The authors found a negative trend at a regional scale, mainly for rays (e.g. *R. clavata*) and sharks (e.g. *Mustelus* spp.).

The overexploitation of sharks has become an urgent Mediterranean ecological issue that requires an international management strategy able to take into account the biological, socio-economic and ethical aspects to preserve the natural equilibrium of the marine ecosystem. Mitigation measures (such as the inclusion of excluder's devices in the trawls or the release of caught or still living specimens) could be introduced for the reduction of fishing impact.

A stronger effort should be sustained to educate fishermen for a responsible activity, as well as collaboration between enterprises and generally among the stakeholders, is highly recommended, also with the goal to define innovative technical solutions.

Acknowledgements

This work was carried out within the Data Collection Regulation and Framework-module trawl surveys MEDITS funded by the European Union and the Italian Ministry of Agricultural, Food and Forestry Policies. We thank all the technical staff of CNR - IAMC UOS of Mazara del Vallo (Italy) involved in data collection and processing.

Author details

Michele Luca Geraci, Sergio Ragonese, Giacomo Norrito, Danilo Scannella, Fabio Falsone and Sergio Vitale*

*Address all correspondence to: sergio.vitale@cnr.it

Institute for Coastal Marine Environment—IAMC National Research Council—CNR, Mazara del Vallo, Italy

References

[1] Vincent ACJ, Hall HJ. The threatened status of marine fishes. Trends in Ecology and Evolution. 1996;**11**:360-361. DOI: 10.1016/0169-5347(96)30041-4

[2] Roberts CM, Hawkins JP. Extinction risk in the sea. Trends in Ecology & Evolution. 1999;**14**(6):241-246. DOI: 10.1016/S0169-5347(98)01584-5

[3] Castro JI, Woodley CM, Brudeck RL. A Preliminary Evaluation of the Status of Shark Species. FAO Fisheries Technical Paper No. 380. Rome: FAO. 1999

[4] Ragonese S, Vitale S, Dimech M, Mazzola S. Abundances of demersal sharks and chimaera from 1994-2009 scientific surveys in the central Mediterranean Sea. PloS one. 2013;**8**(9). DOI: e74865

[5] Lauria V, Gristina M, Attrill MJ, Fiorentino F, Garofalo G. Predictive habitat suitability models to aid conservation of elasmobranch diversity in the central Mediterranean Sea. Scientific Reports. 2015;**5**:13245. DOI: 10.1038/srep13245

[6] Hutchings JA. Influence of population decline, fishing, and spawner variability on the recovery of marine fishes. Journal of Fish Biology. 2001;**59**(sA):306-322. DOI: 10.1006/jfbi.2001.1756

[7] Jennings S, Pinnegar JK, Polunin NV, Warr KJ. Impacts of trawling disturbance on the trophic structure of benthic invertebrate communities. Marine Ecology Progress Series. 2001;**213**:127-142. DOI: 10.3354/meps213127

[8] Arena P. Studio sulla possibilità di razionalizzare e rendere più produttiva la pesca a strascico nel Canale di Sicilia e nel Mediterraneo. Palermo: ESPI. 1985. p. 214

[9] Brander K. Disappearance of common skate, *Raia batis*, from the Irish Sea. Nature. 1981;**290**(5801):48-49. DOI: 10.1038/290048a0

[10] Casey JM, Myers RA. Near extinction of a large, widely distributed fish. Science. 1998;**281**(5377):690-692. DOI: 10.1126/science.281.5377.690

[11] Dulvy NK, Metcalfe JD, Glanville J, Pawson MG, Reynolds JD. Fishery stability, local extinctions, and shifts in community structure in skates. Conservation Biology. 2000;**14**(1):283-293. DOI: 10.1046/j.1523-1739.2000.98540.x

[12] Bradai MN, Saïdi B, Enajjar S. Elasmobranchs of the Mediterranean and Black sea: status, ecology and biology. Bibliographic analysis. In: Studies and Reviews. General Fisheries Commission for the Mediterranean. Vol. 91. Rome: FAO. 2012. p. 103

[13] Malak DA. Overview of the Conservation Status of the Marine Fishes of the Mediterranean Sea. IUCN; 2011

[14] Farrugio H, Oliver P, Biagi F. An overview of the history, knowledge, recent and future research trends in Mediterranean fisheries. Scientia Marina. 1993;**57**(2-3):105-119

[15] Dimech M, Kaiser MJ, Ragonese S, Schembri P. Ecosystem effects of fishing on the continental slope in the Central Mediterranean. Marine Ecology Progress Series. 2012;**449**:41-54. DOI: 10.3354/meps09475

[16] Sarà R. La Pesca a Strascico sui Fondali della Scarpata continentale (Settori di Levanzo e Pantelleria). Rome, Italy: Mem Min Marina Mercantile; 1969. p. 74

[17] Arena P, Li Greci F. Indagine sulle condizioni faunistiche e sui rendimenti di pesca dei fondali batiali della Sicilia occidentale e della bordura settentrionale dei banchi della soglia Siculo-Tunisina. Quad Lab Tecnol Pesca. 1973;**1**:157-201

[18] Macpherson E. Regime alimentaire de *Galeus melastomus* Rafinesque, 1810, *Etmopterus spinax* (L., 1758) et *Scymnorhinus licha* (Bonnaterre, 1788) en Mediterranee occidentale. Vie Milieu. 1980;**30**:139-148

[19] Ragonese S, Di Stefano L, Bianchini ML. Capture and selectivity of cartilaginous fishes in the red shrimp fishery: Strait of Sicily. Biologia Marina Mediterranea. 2000;**7**:400-411

[20] Cingolani N, Coppola SR, Mortera J. Studio di fattibilità per un sistema di rilevazione campionaria delle statistiche della pesca (PESTAT). Quad Ist Ric Pesca Marittima. 1986;**5/l**(2):283-753

[21] Dell'Apa A, Kimmel DG, Clo S. Trends of fish and elasmobranch landings in Italy: associated management implications. ICES Journal of Marine Science. 2012;**69**:1045-1052. DOI: 10.1093/icesjms/fss067

[22] Serena F, Mancusi C, Vacchi M. Threatened species of the Mediterranean Sea. The case of Elasmobranchs fishes: Assessment of their status and international actions for their conservation. In: Gertwagen R, Raicevich S, Fortibuoni T, Giovanardi O, editors. Le interazioni tra uomo ed ambiente nel; Mediterraneo dall'Epoca Romana al XIX secolo: una visione storica ed ecologica delle attivita di pesca. Atti II Workshop Int HMAP Mediterraneo Mar Nero Chioggia, 27-29 settembre 2006; 2008. pp. 127-142. Supplemento ai Quaderni ex ICRAM, 219 p (in Italian with English abstract)

[23] Bombace G, Sara R. La pesca a strascico sui fondali da -500 a -700 metri nel settore a sud-est di Pantelleria. Mem MIN Mar Mercantile. 1972;**33**:63-77

[24] Brčić J, Herrmann B, De Carlo F, Sala A. Selective characteristics of a shark-excluding grid device in a Mediterranean trawl. Fisheries Research. 2015;**172**:352-360. DOI: 10.1016/j.fishres.2015.07.035

[25] Ragonese S, Nardone G, Ottonello D, Gancitano S, Giusto GB, Sinacori G. Distribution and biology of the Blackmouth catshark *Galeus melastomus* Off the southern coasts of Sicily (Central Mediterranean Sea). Marine Science and Engineering. 2009;**10**(1):55-72. DOI: 10.12681/mms.122

[26] Bradai MN, Saidi B, Ghorbel M, Bouain A, Guelorget O, et al. Observations sur les requins du golfe de Gabes (Tunisie meridionale, Mediterranee centrale). Mesogee. 2002;**60**:61-77

[27] Campillo A. Les pecheries francaises de Mediterranee: Synthese des connaissances. RIDRV-92/019-RH Sete. 1992. p. 206

[28] Valls M, Quetglas A, Ordines F, Moranta J. Feeding ecology of demersal elasmobranchs from the shelf and slope off the Balearic Sea (western Mediterranean). Scientia Marina. 2011;**75**(4):633-639. DOI: 10.3989/scimar.2011.75n4633

[29] Scacco U, Andaloro F, Campagnuolo S, Castriota L, Vacchi M. Cartilaginous Fishes as a Component of Trawl Discard in Strait of Sicily. NAFO SCR Doc; 2002, 02/87

[30] Serena F. Field Identification Guide to the Sharks and Rays of the Mediterranean and Black Sea. Rome, Italy: FAO Species Identification Guide for Fishery Purposes; 2005. p. 97

[31] Compagno LJV. Sharks of the World: An Annotated and Illustrated Catalogue of Shark Species Known to Date. Volume 4, Part 1. *Hexanchiformes* to Lamniformes. Rome, Italy: FAO. 1984. DOI: 10.1002/iroh.19860710229

[32] Baino R, Serena F, Ragonese S, Rey J, Rinelli P. Catch composition and abundance of elasmobranchs based on the MEDITS program. Rapport Commission International Mer Méditerranée. 2001;**36**:234

[33] Quero JC Scyliorhinidae. In: Whitehead PJP, Hureau JC, Nielsen J, Tortonese E, editors. Fishes of the Northeastern Atlantic and the Mediterranean. Paris: Unescopp; 1984. pp. 95-100. DOI: 10.1017/s0025315400019640

[34] Bauchot ML. Requins. In: Fischer W, Schneider M, Bauchot ML, editors. Fiches FAO d'identification des especes pour les besoins de la peche Mediterranee et mer Noire, Zone de Peche 37, Revision 1. Rome: FAO; 1987. pp. 767-890

[35] Schembri T, Fergusson JK, Schembri PJ. Revision of the records of sharks and rays species from the Maltese Islands (*Chordata*: *Chondrichthyes*). Cent Medit Nat. 2003;**4**(1):71-104

[36] Serena F, Mancusi C, Ellis J. *Mustelus asterias*. The IUCN Red List of Threatened Species. 2009: e.T39357A10214084. DOI:10.2305/IUCN.UK.2009-2.RLTS.T39357A10214084.en

[37] Fischer W, Bauchot ML, Schneider M. Fiches FAO d'identification des espèces pour les besoins del la péche. Méditerranée er Mer Noire. Zone de Péche 37. Rome, Italy: FAO; 1987

[38] Gracan R, Mladineo I, Lazar B. Insight into the diet composition and gastrointestinal parasite community of the common smooth-hound, *Mustelus mustelus* (*Carcharhiniformes*: *Triakidae*), in the northern Adriatic Sea. Natura Croatica. 2014;**23**(1):35

[39] Saidi B, Mohamed N, Bradai MN, Abderrahman B. Reproductive biology and diet of *Mustelus punctulatus* (Risso, 1826) (*Chondrichthyes*: *Triakidae*) from the Gulf of Gabes, central Mediterranean Sea. Scientia Marina. 2009;**73**:249-258. DOI: 10.3989/scimar.2009.73n2249

[40] Dagit DD, Hareide N, Clo S. *Chimaera monstrosa*. In: IUCN 2012. IUCN Red List of Threatened Species. Iucn website. 2007. Available from: www.iucnredlist.org. [Accessed: 18 October 2012]

[41] Muus BJ, Nielsen JG. Sea fish. In: Muus BJ, Nielsen JG, Dahlstrom P, Nyström BO, editors. Scandinavian Fishing Year Book. Denmark: Hedehusene; 1999. p. 340

[42] Albo PM, Navarro J, Coll M, Aguzzi J, Cardona L, Sáez-Liante R. Feeding ecology and trophic position of three sympatric demersal chondrichthyans in the northwestern Mediterranean. Marine Ecology Progress Series. 2015;**524**:255-268. DOI: 10.3354/meps11188

[43] Torres MA, Ramos, F, Sobrino I. Length–weight relationships of 76 fish species from the Gulf of Cadiz (SW Spain). Fisheries Research. 2012;**127**:171-175. DOI: 10.1016/j.fishres.2012.02.001

[44] Sion L, Bozzano A, D'Ongia G, Capezzuto F, Panza M. *Chondrichthyes* species in deep waters of the Mediterranean Sea. Scientia Marina. 2004;**68**:153-162. DOI: 10.3989/scimar.2004.68s3153

[45] Compagno LJV, Ebert DA, Smale MJ. Guide to the Sharks and Rays of Southern Africa. London: New Holland Ltd.; 1989

[46] Celona A, De Maddalena A, Romeo T. Bluntnose sixgill shark, *Hexanchus griseus* (Bonnaterre, 1788), in the Eastern North Sicilian waters. Boll Mus civ St nat. Venezia. 2005;**56**:137-151

[47] Potoschi A, Iaria G, Spano N. Shark records in the strait of Messina (Central Mediterranean Sea): *Hexanchus griseus* (Bonnaterre, 1788). Rapp Comm Int Mer Medit. 2010;**39**:636

[48] Guallart J, Serena F, Mancusi C, Casper Bm, Burgess GH, Ebert DH, Clarke M, Stenberg C. *Centrophorus granulosus*. In: IUCN 2008. IUCN Red List of Threatened Species. Available from: www.iucnredlist.org

[49] Reiner F. Catálogo dos peixes do arquipélago de Cabo Verde. Publ. Avuls. Inst. Port. Invest. Mar. 1996;**2**:339

[50] Springer VG, Last PR, Stevens JD. Sharks and rays of Australia. Copeia. 1994(4). p. 1055. DOI: 10.2307/1446735

[51] Daley RK, Stevens JD, Last PR, Yearsley GK. Field guide to Australian sharks and rays. CSIRO Div. of Marine Research/Fisheries Research & Development Corporation; 2002. p. 84

[52] Golani D. On deep-water sharks caught off the Mediterranean coast of Israel. Israel Journal of Zoology. 1986;**34**:23-31

[53] Navarro J, López L, Coll M, Barría C, Sáez-Liante R. Short-and long-term importance of small sharks in the diet of the rare deep-sea shark *Dalatias licha*. Marine Biology. 2014;**161**(7):1697-1707. DOI: 10.1007/s00227-014-2454-2

[54] Jones EG, Tselepides A, Bagley PM, Collins MA, Priede IG. Bathymetric distribution of some benthic and benthopelagic species attracted to baited cameras and traps in the deep eastern Mediterranean. Marine Ecology Progress Series. 2003;**251**:75-80. DOI: 10.3354/meps251075

[55] Fanelli E, Rey J, Torres P, Gil de Sola L. Feeding habits of blackmouth catshark Galeus melastomus Rafinesque, 1810 and velvet belly lantern shark *Etmopterus spinax* (Linnaeus, 1758) in the western Mediterranean. Journal of Applied Ichthyology. 2009;**25**(s1):83-93. DOI: 10.1111/j.1439-0426.2008.01112.x

[56] Coelho R, Blasdale T, Mancusi C, Serena F, Guallart J, Ungaro N, Litvinov F, Crozier P, Stenberg C. *Etmopterus spinax*. The IUCN Red List of Threatened Species. 2009. Available from: https://doi.org/10.2305%2Fiucn.uk.2009-2.rlts.t161388a5412576.en

[57] Capapé C. Observations sur le régime alimentaire de 29 Sélaciens pleurotêmes des côtes tunisiennes. Archives de l'Institut Pasteur de Tunis. 1975;**52**:395-414

[58] Capapé C. Diet of the angular rough shark *Oxynotus centrina* (*Chondrichthyes*: Oxynotidae) off the Languedocian coast (southern France, north-western Mediterranean). Vie et Milieu. 2008;**58**:57-61

[59] Guallart J, García-Salinas P, Ahuir-Baraja AE, Guimerans M, Ellis JR, Roche M. Angular roughshark *Oxynotus centrina* (*Squaliformes*: *Oxynotidae*) in captivity feeding exclusively on elasmobranch eggs: An overlooked feeding niche or a matter of individual taste? Journal of Fish Biology. 2015;**87**(4):1072-1079

[60] Serena F, Papakonstantinou C, Relini G, Gil De Sola L, Bertrand JA. Distribution and abundance of Spiny dogfish in the Mediterranean Sea based on the Mediterranean International Trawl Survey program. In: Gallucci V, McFarlane G, Bargmann G, editors. Biology and Management of Dogfish Sharks. American Fishery Society; 2009. p. 139-149

[61] Avsar D. Age, growth, reproduction and feeding of the spurdog (*Squalus acanthias* Linnaeus, 1758) in the South-eastern Black Sea. Estuarine, Coastal and Shelf Science. 2001;**52**(2):269-278. DOI: 10.1006/ecss.2000.0749

[62] Martinho F, Sá C, Falcão J, Cabral HN, and Pardal MÂ. Comparative feeding ecology of two elasmobranch species, *Squalus blainville* and *Scyliorhinus canicula*, off the coast of Portugal. Fishery Bulletin. 2012;**110**(1):71-84

[63] Ebert DA, Serena F, Mancusi C. 2009. *Squalus blainville*. The IUCN Red List of Threatened Species. 2009: e.T161536A5446109. Available from: http://dx.doi.org/10.2305/IUCN.UK.2009-2.RLTS.T161536A5446109.en

[64] Whitehead PJP, Bauchot ML, Hureau JC, Nielsen J, Tortonase E, editors. 1984: Fishes of the North-eastern Atlantic and the Mediterranean. Vol. 1. Paris, France: UNESCO; 1984. p. 510. DOI: 10.1017/s0025315400019640

[65] Ebert DA, Stehmann MF. Sharks, batoids and chimaeras of the North Atlantic. Vol. 7. Roma (Italia): FAO; 2013. p. 523

[66] Ismen A. Age, growth, reproduction and food of common stingray (*Dasyatis pastinaca* L., 1758) in Iskenderun Bay, the eastern Mediterranean. Fisheries Research. 2003;**60**(1):169-176. DOI: 10.1016/s0165-7836(02)00058-9

[67] Notarbartolo di Sciara G, Bianchi I. Guida degli squali e delle razze del Mediterraneo. Muzzio; 1998. p. 388

[68] Lipej L, Mavrič B, Paliska D, Capapé C. Feeding habits of the pelagic stingray *Pteroplatytrygon violacea* (*Chondrichthyes*: Dasyatidae) in the Adriatic Sea. Journal of the Marine Biological Association of the United Kingdom. 2013;**93**(02):285-290. DOI:10.1017/s0025315412000197

[69] Orsi Relini L, Cima C, Garibaldi F, Palandri G, Relini M, Torchia G. La pesca professionale con i palamiti galleggianti nel "Sautuario dei cetacei" del Mar Ligure: si tratta di attivita' ecocompatibili? Biologia Marina Mediterranea. 1999;**6**:100-109

[70] Holtzhausen JA, Ebert DA, Serena F, Mancusi C. *Myliobatis aquila*. International Union for Conservation of Nature. IUCN Red List of Threatened Species, version 2014.3. Available from: www.iucnredlist.org. 2009

[71] Mollet HF, Ezcurra JM, O'Sullivan JB. Captive biology of the pelagic stingray, *Dasyatis violacea* (Bonaparte, 1832). Marine and Freshwater Research. 2002;**53**:531-541. DOI: 10.1071/mf01074

[72] Dulcic J, Lovrenc L, Bonaca M, Jenko R, Grbec B, Guélorget O, Capapé C. The Bull ray in the Adriatic Sea. Cybium. 2008;**32/2**:119-123

[73] Froese R, Pauly D. *Aetomylaeus bovinus* in FishBase January 2016 version. Available from: http://www.fishbase.org/

[74] Wintner S. *Aetomylaeus bovinus*. The IUCN Red List of Threatened Species. 2016. Available from: http://www.fishbase.org/

[75] Steven GA. The British raiidae. Science Progress, London. 1947;**35**:220-236

[76] Yıgın C, Ismen A. Age, growth, reproduction and feed of longnosed skate, *Dipturus oxyrinchus* (Linnaeus, 1758) in Saros Bay, the north Aegean Sea. Journal of Applied Ichthyology. 2010;**26**(6):913-919. DOI: 10.1111/j.1439-0426.2010.01510.x

[77] Serena F, Mancusi C, Barone M. editors. Field guide to the identification of the skates (Rajidae) of the Mediterranean Sea. Guidelines for data collection and analysis. Biologia. Marina Mediterranea. 2010;**17**(Suppl. 2):201

[78] Mulas A, Bellodi A, Cannas R, Cau A, Cuccu D, Marongiu MF, Follesa MC. Diet and feeding behaviour of longnosed skate *Dipturus oxyrinchus*. Journal of Fish Biology. 2015;**86**(1):121-138

[79] Agustin LQ. *Leucoraja circularis*. Sandy Ray. 2009. Available from: http://www.fishbase.org/

[80] Zidowitz H, George M, Fordham S, Kullander SO, Pelczarski W. Sharks in the Baltic: Distribution, Use and Conservation of Cartilaginous Fishes in the Baltic Sea. 2008. Available from: The Shark Alliance. www.sharkalliance.org

[81] Ragonese S, Cigala Fulgosi F, Bianchini ML, Norrito G, Sinacori G. Annotated check-list of the skates (*Chondrichthyes, Rajidae*) in the Strait of Sicily (Central mediterranean). Biologia Marina Mediterranea. 2003;**10**:874-881

[82] Farias I, Figueiredo I, Moura T, Serrano Gordo L, Neves A, Serra-Pereira, B. Diet comparison of four ray species (*Raja clavata, Raja brachyura, Raja montagui* and *Leucoraja naevus*) caught along the Portuguese continental shelf. Aquatic Living Resources. 2006;**19**(2):105-114. DOI: 10.1051/alr:2006010

[83] Bono L, De Ranieri S, Fabiani O, Lenzi C, Mancusi C, Serena F. Studio sull'accrescimento di *Raja asterias* (Delaroche, 1809) (*Chondrichthyes, Rajidae*) attraverso le analisi delle vertebre (Study of the growth of *Raja asterias* (Delaroche, 1809) (*Chondrichthyes, Rajidae*) through the analysis of vertebral sections). Biologica Marina Mediterranea. 2005;**12**:470-474

[84] Navarro J, Coll M, Preminger M, Palomera I. Feeding ecology and trophic position of a Mediterranean endemic ray: Consistency between sexes, maturity stages and seasons. Environmental Biology of Fishes. 2013;**96**(12):1315-1328. DOI: 10.1007/s10641-013-0109-7

[85] Stehmann M, Bürkel DL. *Rajidae*. In: Whitehead PJP, Bauchot ML, Hureau JC, Nielsen J, Tortonese E, editors. Vol. 1. Fishes of the North-eastern Atlantic and Mediterranean. Paris: UNESCO; 1984. pp. 163-196. DOI: 10.1017/s0025315400019640

[86] Gibson C, Valenti SV, Fowler SL, Fordham SV. The Conservation Status of Northeast Atlantic Chondrichthyans; Report of the IUCN Shark Specialist Group Northeast Atlantic Regional Red List Workshop. IUCN SSC Shark Specialist Group; 2006. VIII + p. 76

[87] Kadri H, Marouani S, Saïdi B, Bradai MN, Bouaïn A, Morize E. Age, growth, sexual maturity and reproduction of the thornback ray, *Raja clavata* (L.), of the Gulf of Gabes (south-central Mediterranean Sea). Marine Biology Research. 2014;**10**(4):416-425. DOI: 10.1080/17451000.2013.797584

[88] Kadri H, Marouani S, Bradai MN, and Bouaïn A. Diet and feeding strategy of thornback ray, *Raja clavata* (*Chondrichthyes*: *Rajidae*) from the Gulf of Gabes (Tunisia—Central Mediterranean Sea). Journal of the Marine Biological Association of the United Kingdom. 2014;**94**(7):1509-1516. DOI: 10.1017/s0025315414000587

[89] Ellis JR, Dulvy NK, Serena F. *Raja clavata*. The IUCN Red List of Threatened Species. 2016. Available from: http://www.iucn.org/

[90] Kadri H, Marouani S, Bradai MN, Bouain A, Morize E. Distribution and Morphometric characters of the mediterranean brown ray, *Raja miraletus* (*Chondrichthyes*: *Rajidae*) in the gulf of gabes (Tunisia, Central Mediterranean). American Journal of Agriculture and Forestry. 2014;**2**(2):45-50. DOI: 10.11648/j.ajaf.20140202.15

[91] Follesa MC, Mulas A, Cabiddu S, Porcu C, Deiana AM, and Cau A. Diet and feeding habits of two skate species, *Raja brachyuran* and *Raja miraletus* (*Chondrichthyes, Rajidae*) in Sardinian waters (central-western Mediterranean). Italian Journal of Zoology. 2014;**77**(1):53-60. DOI: 10.1080/11250000802589600

[92] Smale MJ, Ungaro N, Serena F, Dulvy N, Tinti F, Bertozzi M, Mancusi C, Notarbartolo di Sciara G. *Raja miraletus*. The IUCN Red List of Threatened Species. 2009. Available from: http://www.iucn.org/

[93] Bottari T, Rinelli P, Bianchini ML, Ragonese S. Stock identification of *Raja clavata* L. (*Chondrichthyes, Rajidae*) in two contiguous areas of the Mediterranean. Hydrobiologia. 2013;**703**:215-224. DOI: 10.1007/s10750-012-1361-0

[94] Ellis JR, Cruz-Martinez A, Rackham BD, Rogers SI. The distribution of chondrichthyan fishes around the British Isles and implications for conservation. Journal of Northwest Atlantic Fishery Science. 2005;**35**(195-213):113. DOI: 10.2960/J.v35.m485

[95] Relini G, Biagi F, Serena F, Belluscio A, Spedicato MT, Rinelli P, Follesa MC, Piccinetti C, Ungaro N, Sion L, Levi D. I Selaci pescati con lo strascico nei mari italiani. Biol. Mar. Medit. 2000;**7**(1):347-384

[96] Consalvo I, Iraci Sareri D, Bottaro M, Tudisco A, Cantone G, Vacchi M. Diet composition of juveniles of rough ray *Raja radula* (*Chondrichthyes*: *Rajidae*) from the Ionian Sea. Italian Journal of Zoology. 2010;**77**(4):438-442. DOI: 10.1080/11250000903390197

[97] Kadri H, Marouani S, Bradai MN, Boua ÃA, Morize E. Age, Growth, Mortality, longevity and reproductive biology of the white skate, *Rostroraja alba* (*Chondrichthyes*: *Rajidae*) of the Gulf of Gabès (Southern Tunisia, Central Mediterranean). Turkish Journal of Fisheries and Aquatic Sciences. 2014;**14**(1). DOI: 10.4194/1303-2712-v14_1_21

[98] Yiğin CÇ, İsmen A, Umur ÖNAL, Arslan M. Cartilaginous fishes and fisheries in the Aegean Sea. The Aegean Sea. 2015;p. 286

[99] Duman ÖV, Basusta N. Age and growth characteristics of marbled electric ray *Torpedo marmorata* (Risso, 1810) inhabiting Iskenderun Bay, North-eastern Mediterranean Sea. Turkish Journal of Fisheries and Aquatic Sciences. 2013;**13**(3). DOI: 10.4194/1303-2712-v13_3_19

[100] Capapé C, Crouzet S, Clement C, Vergne Y, Guelorget O. Diet of the marbled electric ray *Torpedo marmorata* (*Chondrichthyes*: *Torpedinidae*) off the Languedocian coast (Southern France, Northern Mediterranean). ANNALES Series Historia Naturalis. 2007;**17**:17-22

[101] Consalvo I, Scacco U, Romanelli M, Vacchi M. Comparative study on the reproductive biology of *Torpedo torpedo* (Linnaeus, 1758) and *T. marmorata* (Risso, 1810) in the central Mediterranean Sea. Scientia Marina. 2007;**71**(2):213-222. DOI: 10.3989/scimar.2007.71n2213

[102] Romanelli M, Consalvo I, Vacchi M, Finoia G. Diet of *Torpedo torpedo* and *Torpedo marmorata* in a coastal area of Central Western Italy (Mediterranean Sea). Marine Life. 2006;**16**:21-30

[103] Abdel-Aziz SH. Observations on the biology of the common torpedo (*Torpedo torpedo*, Linnaeus, 1758) and marbled electric ray (*Torpedo marmorata*, Risso, 1810) from Egyptian Mediterranean waters. Marine and Freshwater Research. 1994;**45**(4):693-704. DOI: 10.1071/mf9940693

[104] Marouani S, Chaâba R, Kadri H, Saidi B, Bouain A, et al. Taxonomic research on *Squalus megalops* (Macleay, 1881) and *Squalus blainvillei* (Risso, 1827) (*Chondrichthyes*: *Squalidae*) in Tunisian waters (central Mediterranean Sea). Scientia Marina. 2012;**76**(1):97-109. DOI: 10.3989/scimar.03440.22b

[105] Iglésias SP. Chondrichtyens du Nord-est Atlantique et de la Méditerranée, (Une classification naturelle basée sur des spécimens de collection, avec barcodes DNA et photographies standardisées), Volume I (plates), Provisional Version 06, 01 avril 2012. p. 83. Available from: http://www.mnhn.fr/iccanam/ [Accessed: 23 November 2012]

[106] Russo T, Parisi A, Garofalo G, Gristina M, Cataudella S, Fiorentino F. SMART: A spatially explicit bio-economic model for assessing and managing demersal fisheries, with an application to Italian trawlers in the strait of Sicily. PLoS ONE. 2014;**9**(1):e86222. DOI: 10.1371/journal.pone.0086222

[107] Gancitano V, Enea M, Colloca F, Gancitano S, Ingrande G, Massi D, Rizzo P, Titone A, Fiorentino F. Temporal dynamics of demersal resources in the south of sicily (GSA 16) during the last twenty years. Biologica Marina Mediterranea. 2015;**1**. ISSN 1123-4245

Chapter 3

Fisheries Bycatch of Chondrichthyes

Sara Bonanomi, Jure Brčić', Alessandro Colombelli, Emilio Notti, Jacopo Pulcinella and Antonello Sala

Additional information is available at the end of the chapter

http://dx.doi.org/10.5772/intechopen.69334

Abstract

This chapter reviews the current state of knowledge about global chondrichthyes bycatch, including how the combination of biological characteristics of these species and the impact of different fishing methods reflect their vulnerability to bycatch. Specifically, it summarizes the present status of elasmobranchs and chimaeras bycatch worldwide. Following this, it illustrates the main ecological consequences behind such incidental captures. Hence, it provides a description of how chondrichthyes bycatch occurs across broad fishing gear categories. Moreover, it outlines potential mitigation options available to reduce bycatch. As such, the main technical measures tested and applied in different fishing practices are listed. In addition, an overview about which are the current management measures implemented at international level to ensure long-term conservation of cartilaginous fish is reported. Then, two case studies summarize how chondrichthyes bycatch occurred in some of the most exploited fishing areas. Finally, the chapter ends with relevant considerations about potential for new research needed to monitor and reduce chondrichthyes bycatch.

Keywords: bycatch, fisheries, chondrichthyes, fishing gears, mitigation measures

1. Introduction

The unintentional catch of non-target species or bycatch occurring in fisheries represents one of the major threats to long-lived marine species worldwide. In this context, chondrichthyes are thought to be particularly vulnerable to bycatch due to their biological characteristics that include long life spans, late age at maturity, larger size at birth and low reproductive rates [1–3]. Historically, despite the high value of some cartilaginous fish products (e.g. shark fin market), sharks and their relatives have often been considered as low-value marketed fish, mainly landed as bycatch of other more profitable species (e.g. tuna and tuna-like species [4, 5]). Due to the high level of exploitation, they are now considered among the most threatened marine animals worldwide. Indeed, based on The International Union for Conservation of Nature (IUCN) Red List categories and modelled estimates, of the 1041 assessed species, at least 24% are predicted to be threatened worldwide [6]. According to the most recent the Food and Agriculture Organization of the United Nations (FAO) report [7], approximately 766,064 tons of chondrichthyes were caught in 2011, mainly sharks and batoids. A large portion of these catches was a result of unintentional catches. In addition, illegal and unregulated fishing activities might have increased the total catch because some cartilaginous fish are often discarded at sea without being recorded in any assessment [8–11]. Thus, fisheries bycatch is likely to have significant demographic impacts on these species. Therefore, there is an urgent need to better understand the interactions between chondrichthyes and fisheries in order to develop and apply proper tailored fisheries management strategies.

1.1. Elasmobranchs bycatch

Elasmobranchs are often taken as bycatch in fisheries targeting more valuable species (e.g. tuna and tuna-like species [4, 5]), although in some countries they are caught as target species (e.g. blue shark in the North East Atlantic). Over the last century, as a result of overexploitation, severe population declines have been documented for a number of non-target species mainly large pelagic sharks in the Mediterranean (**Figure 1** and [12, 13]) and North West Pacific [1, 10]; skates and sawfish in the Atlantic [6, 14, 15]. Such declines have been related either to a direct fishery and/or indirectly to incidental capture. According to the Food and Agriculture Organization of the United Nations (FAO), since the 1950s, landings of sharks and rays gradually increased and peaked in 2003 with more than 400,000 tons, afterwards declined by 20% [6, 7]. In contrast, recent investigations estimated a potential global catch rate of about 100 million sharks in 2000 and between 63 and 273 million sharks per year [16]. In any case, over the last decades, elasmobranch catches have been gradually dominated by rays. Due to massive exploitation and their relatively large body size, rays and skates are among the most threatened elasmobranch [6]. While several species are subjected to targeted fishing, elasmobranch catches are often illegal, unregulated, discarded and unreported to national or international management agencies. In addition, it has been estimated that approximately 50% of the global elasmobranch catches consists of individuals incidentally captured mostly in pelagic longline fisheries [11, 17]. Nevertheless, these catches still remain poorly documented. Thus, nominal annual landings do not reflect the true magnitude of fisheries-related mortality. Bycatch of sharks and rays results in a consistent number of individuals often discarded dead or dying at sea. Hence, the real number of elasmobranch discarded has generally not been reported. Based on the initially available estimates, approximately 27.0 million tons of individuals have been discarded between 1980 and 1993 [18] and about 7.3 million tons between 1998 and 2001 [4] with large uncertainties. Currently, some nations have invested in research and monitoring programmes, evaluating bycatch and discard by trained fisheries observers during commercial fishing operations [19–21]. Overall, a number of studies revealed different pattern in bycatch among gears, fisheries, fishing areas and species composition [10, 11, 22] and references there in. For sharks, pelagic longlines targeting tuna and tuna-like species reflect the highest global shark bycatch ratio and one of the largest discard ratio [23, 24].

Figure 1. Common thresher shark (*Alopias vulpinus*) caught as bycatch by midwater pair trawling in the Adriatic Sea (Mediterranean Sea). Photo credit: Andrea Petetta.

Notably, blue shark (*Prionace glauca*) is the most encountered bycatch species in pelagic longline fisheries mainly around sub-tropical and temperate regions [5, 25]. This species is currently listed as Near Threatened globally, while as Critically Endangered in Europe and Mediterranean [26]. According to recent literature, this species together with other pelagic sharks is now considered a second target species in several pelagic fisheries (e.g. in the North East Atlantic [27]) and in countries where management and conservation measures are very limited (e.g. pelagic fisheries supplying shark fin in some Pacific islands [28, 29]). In contrast, for rays, trawl fisheries seem to pose the highest threat in terms of bycatch [11, 30, 31] while longline and gillnet fisheries show the highest discard ratio [11, 32]. Of particular note, in coastal trawl fisheries, rays comprise a substantial component of the bycatch and some species might be retained and sold [6, 33]. For instance, the thornback skate (*Raja clavata*) is a target species of most of inshore trawl fisheries in Europe [34]. It also dominates bycatch in offshore mixed trawl fisheries particularly targeting flat-fishes and cod in the North Atlantic [34–36]. In these fisheries, a considerable number of individuals (mainly juveniles and small individuals) are generally discarded and are likely to be dead. Hence, this species has been globally and locally listed as Near Threatened [37]. Overall, there is a wide spread concern about the impact of trawl fisheries on elasmobranch in general. Given their large diversity, several species constitute considerable portions of trawl bycatch because a viable market exists for them [6, 11, 22]. In addition, misidentification is quite common in multi-species trawl fisheries, especially for those species which are partially or not target species (i.e. rays and skates) where only little information on post-capture mortality is available.

1.2. Chimaeras bycatch

Despite one species (i.e. the elephant fish *Callorhinchus milii* in New Zealand [38]), chimaeras do not seem the subject to any direct fishing pressure. Given the minimal fishery interest and very limited reported catches, official landing statistics of chimaeras are generally gather together with sharks and rays. Overall, chimaeras are occasionally taken as bycatch by deep-water trawl fisheries and by deep-water longline, and most likely discarded at sea across their geographic distribution (**Figure 2**; [39]). The occurrence of several chimaeras species in deep-water habitats and the continuous expansion of deep-water trawl fisheries raised awareness about their potential overexploitation. Although for chimaeras, very little information exists on their ecology, life history traits and behaviour; there is a growing concern about their vulnerability to population depletion [40, 41]. Like other chondrichthyes, the extent of chimaeras bycatch is poorly documented and individuals are usually discarded at sea due to scarce commercial value. For instance, the rabbitfish *Chimaera monstrosa* is probably one of the most known chimaeroid fishes. This species is often retained as bycatch in the North East Atlantic and Mediterranean Sea [41]. Even though no specific data on population trend over time are available, this species is assessed as Near Threatened due to suspected but unconfirmed declines [41, 42]. Notably, discard is a critical aspect for this species and its relatives. The soft body structure of chimaeras and the depth of capture decrease the chance of survival of individuals after being discarded.

Figure 2. Small-eyed rabbitfish (*Hydrolagus affinis*) caught as bycatch by deep-water longline in the North East Atlantic (Azores). Photo credit: Sara Bonanomi.

http://dx.doi.org/10.5772/intechopen.69334

Evidence of low post-capture survival suggests that rabbitfish may represent at least 10% of discarded biomass in deep-water trawl fisheries off West coast of Ireland [40]. However, due to the lack of biological information available for these species, further research is needed in order to understand the real impact of incidental captures of chimaeras across their habitats, which are the ecological implications.

2. Understanding the impact of fisheries bycatch on chondrichthyes

Cartilaginous fish are particularly vulnerable to overexploitation since they are slow-growing K-selected species. Evidence of their vulnerability is confirmed by the fact that poor data of sustainable target and non-target fisheries exist and only few countries have adopted specific technical strategies and management plans for these species. Overall, vulnerability to fisheries bycatch reflects intrinsic life-history traits and population parameters of chondrichthyes correlated to fisheries interactions. In this view, the following paragraphs summarize the ecological consequences beyond bycatch of chondrichthyes and the difference in species composition among different gear types.

2.1. Ecological consequences

Chondrichthyes are among the most wide-ranging apex predators and key actors in structuring ocean ecosystems [13, 43]. The incidental capture of these top predators is a result of indirect harvesting, which negatively affects trophic interactions in the marine food webs. The indirect removal of sharks and their relatives cause a number of changes in community structure and food chain such as between prey-predator and predator-competitor interactions, top-down control structuring, species replacement, increasing discard rates and scavengers density [1, 13, 44]. Several studies have demonstrated that fisheries bycatch poses a serious threat to the long-term survival of chondrichthyes in general. Due to specific life-history traits, the survival and population size of these species might be severely compromised after intense extirpation. The ability of cartilaginous fish to recover after depletion is low on average. However, incidental capture does not affect all chondrichthyes similarly. Differences in terms of long-term survival, reproductive fitness and resiliencies have been documented. Generally, elasmobranchs exhibit a relatively higher tolerance to mortality associated with bycatch compared to chimaeras. The latter together with deep sharks are the most affected and vulnerable chondrichthyes. The mortality of deep-water species is assumed to be high, given their physical characteristics (soft-bodied) and increasing physiological stress, once individuals are unintentionally caught and then immediately discarded at sea [45, 46]. Still little is known on the discard survival of these species in most fisheries. In contrast, most scientific literature is focused on large pelagic sharks. These species are suffering bycatch in many fisheries and are experiencing dramatic population declines across their geographic distribution. Of particular note, the loss of large pelagic sharks has often resulted in complex community changes and trophic cascades in the marine food web, with a consequent increase of high-level pelagic elasmobranch mesopredators. It has been documented that as commercial fisheries develop and increase, the reduction of large pelagic sharks has been observed and corresponded with an increasing trend of mesopredator like smaller sharks and rays [47, 48]. The reduction of large pelagic predators influences the abundance, distribution and behaviour of smaller long-live species, including sharks and rays. It also contributes to a decreasing natural mortality of preys having few other predators and competitor species. For instance, evidence of significant declines in large pelagic predators has led to increasing expansion and abundance of pelagic stingray (*Pteroplatytrygon violacea*) in the tropical Pacific [49]. While, other studies showed that a simultaneous removal of large sharks and elasmobranch prey species would not affect mesopredator abundance in case these species are a significant prey source for top predators [50, 51]. In addition, like other chondrichthyes, large pelagic sharks might have different species-specific bycatch survival. As previously mentioned, blue shark (*Prionace glauca*) is a major bycatch species in pelagic longline fisheries targeting tunas and billfish worldwide and it can reach up to 70% of the total catches [52, 53]. Even if this species is intensively exploited, it exhibits faster reproductive potential compared to other pelagic sharks like shortfin mako (*Isurus oxyrinchus*) and porbeagle (*Lamna nasus*) incidentally taken. Having lower fecundity and higher age at maturity, these species would appear more susceptible to bycatch mortality and would be less capable of population changes if caught in substantial numbers [48, 54]. Moreover, comparing different chondrichthyes, resilience is higher for small coastal species rather than pelagic and low for large coastal and deep-water species [55]. Given their intrinsic life-history traits, juvenile survival rather than fecundity is a crucial factor contributing to population growth rate, especially in longer-lived sharks [2, 56]. Thus, in several sharks and relatives both growth and juvenile survival might be severely affected by indirect fishing pressure.

2.2. Differences among fishing gears

Bycatch takes place in several fishing practices across the entire chondrichthyes geographic distribution. Different interactions among gear type and species composition varied according to the magnitude of fishing effort and gear selectivity. Although a multitude of gear types exists, chondrichthyes bycatch mainly occurred in four broad categories of fishing gears.

2.2.1. Longline

Longline constitutes one of the major sources of individual-chondrichthyes bycatch worldwide. Due to specific technical characteristics, this type of gear catches a significant amount of non-target species mainly cartilaginous fish. Longline comprises a main line carrying a number of hooks (between 100 and 300) on branch lines (snoods) of variable lengths (up to 9 km) and spacing. It may be set either at or near the bottom (bottom-set longline), drifting in mid-water, or near the surface (pelagic surface longline [57]). Bottom longline is generally used to catch demersal and/or deep fish (cod like, flatfish), while pelagic longline is commonly targeting large pelagic fish (tuna, swordfish and billfish). Species composition and selectivity can be affected by hook (shape, size, gap, etc.) and bait type in the longline system. Chondrichthyes bycatch may strongly increase by using smaller hooks but effects could be species-specific [58] (see Section 3). In addition, the great depths that longline can reach, long hauling and soak time severely influence bycatch mortality. Deep-water sharks and chimaeras are frequently caught as bycatch by deep-water longline and they are immediately discarded dead at sea.

However, the portion of chondrichthyes bycatch in pelagic longline fisheries is the greatest so far. Large pelagic sharks account for more than 50% of the chondrichthyes bycatch in pelagic longline fisheries [25, 59], with blue shark accounting for the largest catches in temperate and sub-tropical areas and silky shark in the tropics, instead [5, 11].

2.2.2. *Gillnet*

Gillnets can entangle a variety of chondrichthyes, mainly demersal sharks and rays, in both direct and indirect fisheries. A gillnets is a single netting wall held vertically in the water column by floats on the headrope and weights on the footrope [57]. This type of gear can be held on the sea surface or at certain depth by floats to drift (driftnet), or set on the sea floor and anchored by weights (bottom gillnet). Pelagic fish, including marketed sharks, are usually caught by driftnets, while groundfish are commonly targeted by bottom gillnets. Generally, depending on the mesh size and target species, gillnets are highly selective gears. However, a number of non-target and vulnerable species can become entangled around the head or mouth as they try to escape. For chondrichthyes, the highest bycatch rate inferred by gillnets has been reported in the North Atlantic [11]. Generally, sharks and rays are the most encountered species [32]. Among sharks, Carcharhinidae and Sphyrnidae species seem to have the highest mortality rates [35], while less information on ray bycatch in gillnet fishery is available given that these species are usually discarded alive and sometimes retained on board [11].

2.2.3. *Trawl*

Trawl is considered the most detrimental fishing gear for numerous elasmobranches and chimaeras [60]. Several trawl fisheries exist depending on gear-species interactions and fishing area. Generally, trawl consists of one funnel shaped net either towed by one or two boats in midwater (pelagic trawling) or along the sea bottom (bottom trawling). Pelagic trawling usually target small pelagic fish (anchovies and sardine), while bottom trawling mainly catch groundfish, shellfish and shrimps. A number of factors, such as duration of the trawl, the size of the catch and sorting time can influence post-capture survival of vulnerable species like chondrichthyes [60]. Of particular note, in a contest of multi-species trawl fishery (e.g. coastal trawl fisheries in the Mediterranean), bottom trawling can cause a substantial impact by changing demersal fish assemblages, thus reducing chondrichthyes abundance and diversity [61]. Several non-target elasmobranches and chimaeras are often discarded after being caught by bottom trawl fisheries in many areas [60]. The North West Pacific, North East Atlantic and Mediterranean Sea have historically been the most exploited areas by trawl fisheries with a consequent high rate of chondrichthyes bycatch [11, 13, 62]. Recently, the trend has expanded in the Central Pacific and Southern Atlantic, but little information has been provided due to unassessed fisheries or poor research.

2.2.4. *Purse seine*

Compared to previous fishing gears, the interactions between chondrichthyes and purse seine are apparently limited. This type of gear is usually used to target schooling of small (anchovies and sardine) and large pelagic fish (tuna and tuna like species). However, during the detection of fish schools and encircling operations, non-target species mainly large predators attracted by high fish density might be retained within the net. Purse seine fisheries are recently posing an increasing indirect removal of large pelagic sharks in tropical and subtropical areas in the Indian and Pacific oceans [53]. In these regions, a low level of shark bycatch has been reported with a majority of juvenile of silky shark being caught [25, 63]. Nevertheless, given the growing annual tuna production in the Pacific, estimates of shark bycatch registered for purse seine are likely to increase in the near future [5, 64].

3. Mitigation options to reduce chondrichthyes bycatch

Understanding the post-capture mortality of cartilaginous fish bycatch can provide insights into their intrinsic biological traits driving mortality. This knowledge is the basis for develop and introduce management measures and technical options in order to mitigate lethal effects of capture [60, 65].

3.1. Conservation and management measures

Threatened species like chondrichthyes would significantly benefit from conservation and management measures implemented synergistically at international and national levels [66, 67]. Historically, international efforts devoted to the conservation and management of sharks have been poorly pursued [60, 68]. During the last decades, several attempts have been made at national or local levels in response to specific conservation objectives for certain species and habitats [66–68]. However, the co-operation among national authorities to encourage regional and global measures has been generally inadequate and/or inefficient. This inconsistency and this lack of uniformity mean that, today at the international level, the protection and conservation of the species of cartilaginous fish are entrusted to a set of laws that often result fragmented, duplicated or overlapping, leaving room for weaknesses and gaps. Traditionally, there have not been specific laws tailored precisely to address the bycatch phenomenon, but instead the limitation of bycatch has been one of the several aspects taken into account into wider frameworks dedicated to the management and conservation of the world's aquatic living resources [68]. However, different approaches to shark conservation and management exist, and among the international laws that protect shark species, it is easy to distinguish between fishery regulations, conservation approaches (based on listing of specific species) and habitat protection. For instance, The United Nations Convention on the Law of the Sea (UNCLOS) asks to countries members to co-operate directly or through international organizations in relation to highly migratory species (UNCLOS Articles 63 and 64). This obligation is relevant to many oceanic shark species that are listed as highly migratory species in Annex I of the United Nations Convention Law of the Sea. A particular attention is also devoted to addressing incidental issues related to bycatch. In addition, the Port State Measures Agreement (PSMA) prevents that illegal and unreported fishing, as well as bycatch caught fish, may enter international markets through ports. Since 2000, the international trade in cartilaginous fish has been regulated by The Convention on the International Trade in Endangered Species and Wild (CITES), and currently, seven species are listed in

Appendix I (endangered species, whose trade International is permitted only in exceptional cases), and 11 species are included in Appendix II (species not necessarily threatened with extinction, but whose international trade must be controlled in order to avoid an exploitation incompatible with their survival). The Appendix III may include species not necessarily threatened with extinction at the global level, and their inclusion may be decided without the agreement of other State parties, in order to generate concern on a particular species and open the door to international co-operation. Beyond the practical effects of including shark species in the CITES Appendices, the parties have repeatedly recommended improved interventions for the conservation and management of sharks, inviting the FAO Committee of Fisheries (COFI) and the regional commissions for the management of Fishing to intensify efforts, to undertake research, training, data collection and data analysis and to indicate as a priority the development of action plans for the conservation and management of sharks. Furthermore, sharks and relatives are a focus group of the Convention on Biological Diversity (CBD), which plays a critical role in habitat protection, identifying the correct strategies for both *in situ* and *ex situ* conservation, as well as delivering solid guidelines necessary for protected areas establishment [69]. Together with the management and conservation measures comes the International Plan of Action (IPOA) for the conservation and management of sharks, which it has been meant to be received as a set of technical directives for the conservation and management of sharks, to be used as guidelines by FAO member countries during the implementation of country-specific National Plans of Action. When the plan was firstly adopted in 1999, the overall knowledge available about the fishery status of shark in the world was utterly scarce and only a few countries had specific plans including fishing regulations of cartilaginous fish. Given the wide distribution of these species, it became clearer that international co-operation was a key aspect for conservation and protection of cartilaginous fish populations. The IPOA-Shark aims mainly at the conservation of sharks and at their long-term and sustainable use. The plan applies to all cartilaginous fish, including sharks, rays, skates and chimaeras and to all kind of fisheries, including commercial catches, recreational fishing, direct fishing or by-catch and others. The IPOA-Shark suggests the use of many well-known strategies of biodiversity conservation and fishery management, including a better control over access of fishing vessels to shark stocks, a decrease in fishing effort in any shark species where catch is unsustainable, the identification of vulnerable and threatened species, an improved data collection, the assessment and reporting, a sustainable use of target species and a full utilisation of dead sharks, research on little known shark species and gathering of utilization and trade data on shark species [70]. The Plan establishes not only the need to manage shark catches, but also calls for a better legislation on incidental catches, particularly in the tuna fishing sector. The plan also recommends that States that implement the Shark-plan, at least every 4 years should assess its implementation for the purpose of identifying cost-effective strategies for increasing its effectiveness, while states that determine that a Shark-plan is not necessary should review that decision on a regular basis taking into account changes in their fisheries. Finally, the IPOA-Sharks address shark conservation and management in a more comprehensive way than is achieved in the treaties previously discussed in this chapter, but one of its core problems, from a legal perspective, is that it does not create binding rights and obligations on states, because it is not a treaty or a 'hard' law [68].

3.2. Technical measures applied to different fishing gears

Table 1 summarize the current technical measures adopted or proposed in different fishing gears in order to prevent and reduce chondrichthyes bycatch.

Gear type	Technical measure	References
Pelagic longline	Circle hooks	[24, 71–78]
	Bait restrictions	[71, 73, 79, 80]
	Bans on wire leaders	[81–84]
	Hook depth	[49, 85, 86]
	Temperature avoidance	[48, 87, 88]
	Reducing soak time	[4, 87, 88]
	Repellents	[89–91]
Bottom longline	Number of hooks	[92, 93]
Gillnet	Mesh size regulations	[94, 95]
	Tensioning gillnet	[96]
Trawl	Bycatch reduction devices	[97, 98]
	Filter grid	[99, 100]
Purse seine	Ecological FADs	[101, 102]
	Deterrents	[103]
	Times	[104]
	Restriction on sets on FADs and other floating objects	[105]
	Multiple FADs	[102]

Table 1. Gear types and associated technical measures.

4. Case studies

Despite indirect fisheries removal is a concern for several long-lived marine species, chondrichthyes bycatch is a noteworthy complex issue to investigate and manage. The following paragraphs summarize the current knowledge of chondrichthyes bycatch and strategies for reduction in areas where most of the research is carried out.

4.1. Chondrichthyes bycatch in the North Atlantic and potential solutions

Most of the information concerned about chondrichthyes bycatch has been reported for North Atlantic region. In this area, sharks and their relatives are both directly and indirectly taken by numerous commercial fisheries mainly tuna and billfish longline, driftnets, purse seines and shrimp trawls. Considering the North East Atlantic, spurdog (*Squalus acanthias*) constitutes a significant bycatch component of direct fisheries, mainly in shrimp trawlers and gillnet fisheries [34, 106]. In addition, a substantial number of spurdog are often discarded at sea with a relative low mortality (6%) reported for trawls [107, 108] and higher levels (55%) for gillnets [108] in this area. However, given that this species has a long history of exploitation in the North East Atlantic region; recent restrictive management measures have been adopted. For instance, only a small bycatch quota is allowed in Norwegian waters (no more than 5% by live weight of the catch retained on board and bycatch must not exceed 20% of the total landings in certain periods [34]). While, direct fishing with nets and longline has been prohibited in Swedish waters since early 2008 and effort restrictions have been applied in the mixed fisheries in the North Sea, West of Scotland and Irish Sea instead (EC 1342/2008 [34]). Furthermore, several deep-water sharks, skates and chimaeras are often taken as bycatch in mixed fisheries in the North East Atlantic. For instance, leafscale gulper shark (*Centrophorus squamosus*) and Portuguese dogfish (*Centroscymnus coelolepis*) are incidentally captured and discarded in the longline fisheries targeting black scabbardfish (*Aphanopus carbo*) in Portugal, Azores and Madeira islands [109–111]. Both species have been recently assessed as Endangered in the Red List of European marine fish [112]. According to European regulations, no directed trawl and gillnet fisheries are permitted for several deep-water sharks species in certain areas (see EC No 1568/2005 and EC No 41/2007 [34]). Only a small fraction of bycatch quota is allowed (between 3 and 10%) based on gear type and fishing area [34]. However, given that black scabbardfish and deep-water sharks apparently spatially overlap, other management measures should be introduced [109]. In contrast, many ray and skate species are usually taken as target species and bycatch in demersal fisheries for flatfish with thornback ray (*Raja clavata*) and thorny skate (*Amblyraja radiata*) accounting for 90–95% of the total skate bycatch in the North Sea and in the Barents Sea ecoregions, respectively [113, 114]. In the North West Atlantic, sharks and their relatives are generally taken in direct fisheries and also incidentally captured, mainly in USA, Canada and Mexico (Atlantic side), and with less extent in Caribbean waters [16, 115, 116]. In USA and Canada, large pelagic sharks are intensively harvested by longline fisheries. In addition, these species are the most encountered in reliable commercial landings and a large portion of these is a result of indirect fishing removal [52, 53]. While, in Mexico and Caribbean waters, sharks and rays are usually fished by artisanal fishermen, and juveniles of large species are often taken as bycatch [117]. However, regional management and conservation measures have been implemented in some fisheries. For instance, the United States National Marine Fisheries Service (NMFS) in USA and the Department of Fisheries and Oceans (DFO) in Canada have conducted observer programs to monitor commercial fishing vessels targeting tuna, swordfish and sharks. In addition, other measures including the reduction of fishing effort targeting sharks, time/area closure in certain seasons and gear restrictions have been adopted in the North Atlantic [116, 118–120]. Notably, many large pelagic sharks are overfished or in overexploitation and are often removed as bycatch in longline fisheries targeting tuna and swordfish in the North Atlantic as a whole [11, 54, 117]. Given their wide distribution and their high ability to migrate, large pelagic sharks have been usually exploited by different nations in the North Atlantic. Hence, blue shark (*Prionace glauca*), shortfin mako (*Isurus oxyrinchus*) and porbeagle (*Lamna nasus*) are among the most common caught species in the area, with blue shark having the highest bycatch and discard rates [11]. Recently, specific technical mitigation measures have been proposed and implemented in order to reduce bycatch mortality. Some authors suggested the use of circle hooks rather than J-hooks to increase post-release survival of large pelagic sharks in longline fisheries [71, 72, 120].

Overall, monitoring and management of cartilaginous fish in the North Atlantic are covered by different international advisory parties like International Commission for the Conservation of Atlantic Tunas (ICCAT), North West Atlantic Fisheries Organisation (NAFO), Inter-American Tropical Tuna Commission (IATTC). Such organisations have recently recommended a number of measures concerning the reduction of fishing mortality in fisheries directly and indirectly targeting endangered or particularly vulnerable large pelagic sharks, technical gear modifications to increase fishing selectivity, monitoring the retaining of shark fins, bycatch and discard by appropriated scientific observer programs and reporting of catch data [53].

4.2. Chondrichthyes bycatch in Mediterranean Sea and potential solutions

Although rarely targeted in Mediterranean fisheries, chondrichthyans are often caught as bycatch [121, 122]. According to [119], in the Eastern and Southern Mediterranean countries some chondrichthyan bycatch species provide an important source of cheap fishmeal. A long history of trawling in Mediterranean resulted in significant loss in number of recorded elasmobranch since 1950s [13]. Mediterranean trawls are mostly equipped with either 40 or 50 mm diamond or 40 mm square mesh codends [123]. These codends are able to release a certain amount of small chondrichthyan species [124, 125], but are certainly far from optimal. It is difficult to expect that there is going to be any further increase in mesh size in Mediterranean in near future, which is why some authors advocate using various excluder grid devices, e.g. Turtle Excluder Device (TED) [60, 66, 67, 126, 127]. So far, only one study examined the effect of inserting an excluder grid device in front of the size selective codend with the aim to reduce shark bycatch in the Western Mediterranean bottom trawl fishery [60]. The authors tested one grid with 90 mm grid bar spacing, and the results showed that the grid was not very effective in excluding *Galeus melastomus* (the only shark species tested in the study), and that reduction from 90 to 70 mm bar spacing would be much better compromise in excluding this species while maintaining the catches of commercial species. Moreover, the study performed by [128] summarized the results from the monitoring programme of accidental catches of cetaceans by Italian midwater trawlers, where highly vulnerable elasmobranch bycatch species were also monitored. According to their study, a total of 15 different species of elasmobranchs were caught in the Adriatic Sea midwater trawl fishery between July 2006 and December 2008. Sharks and rays, although not targeted, by this fishery, are highly valued and were regularly marketed. Some shark species like thintail threshers (*Alopias vulpinus*), piked dogfish (*Squalus acanthias*) and smooth hounds (*Mustelus mustelus*) were caught in relatively large numbers. Only pelagic stingrays (*Pteroplatytrygon violacea*), common stingrays (*Dasyatis pastinaca*), bull rays (*Aetomylaeus bovinus*) and common eagle rays (*Myliobatis aquila*) were regularly discarded [128]. Since the study showed seasonal and geographical variation in elasmobranch bycatch rate, the authors argued that if any mitigation measures are to be implemented in the study area, both geographical and seasonal variation must be taken into account. They also identified the North Adriatic Sea as an area where mitigation measures should be adapted first. Ref. [50] focused only on the bycatch of myliobatidae rays, common eagle ray (*Myliobatis aquila*) and bull ray (*Aetomylaeus bovinus*) in the midwater trawl fishery in the Mediterranean Sea. Bull ray catch data from the Mediterranean region are very scarce; IUCN listed this species under Data Deficient category [129]; but nonetheless, due to its life history traits it can be considered as Potentially threatened locally [50, 130]. According to [130] and references therein, the greatest threats to bull ray in Greek waters (Eastern Mediterranean) are gillnets where they are caught as a bycatch and often discarded [130]. IUCN listed common eagle ray also as data deficient, but from [131], we know that commercial demersal catches have dramatically declined in Gulf of Lion during the 1970–1995 period, what can probably be extrapolated to other Mediterranean regions. Ref. [50] noticed that the increase in the haul duration in Northern Adriatic midwater pair trawl fishery resulted in lower CPUE values for these two species, probably due to very low likelihood of catching more than few individuals per haul. They further concluded that limitation on haul duration and midwater trawl size could be beneficial, since, once discarded, it can greatly increase its chance of survival. Elasmobranch bycatch is also known to be relatively high in longline fishery [17]. A surface longline fishery targeting tuna and swordfish in the Mediterranean often incidentally captures several pelagic sharks [132]. According to [133], shark bycatch is higher in the Western than in the Eastern Mediterranean Sea, where sharks were reported to be the second most important bycatch group caught with traditional swordfish, American type and albacore longline types. The blue shark, *Prionace glauca,* was the most abundant shark bycatch species, followed by tope shark (*Galeorhinus galeus*), shortfin mako (*Isurus oxyrinchus*) and common thresher shark (*Alopias vulpinus*). The authors concluded that in any drifting longline fishery targeting large pelagic fish, many hundreds or thousands of sharks are killed in the Eastern Mediterranean each year [133]. Bottom longlines are typically catching demersal chondrichthyan species such as *Mustelus* spp., *Squalus* spp., *Torpedo* spp., *Galeus melastomus, Hexanchus griseus, Centrophorus granulosus* and several ray species such as *Raja clavata* and *Raja radula* [134–136]. Ref. [126] argued that a part of the solution for elasmobranch bycatch reduction in bottom or near bottom longlines could be the adoption of nylon gangions as opposed to the steel wires because this would allow the larger specimens to bite of the line and escape. This, however, would not reduce the bycatch of juveniles. Several technical solutions could increase the selectivity of both bottom and surface longlines, e.g. hook size and shape, reduced soaking time, fishing depth, avoidance of certain types of baits, presence of swivels, etc. [126, 127].

5. Conclusion

Chondrichthyes are among the most threatened long-lived marine species worldwide. Incidental capture is one of the main sources of mortality for these species. However, prevention and reduction of chondrichthyes bycatch are not easy tasks. Ongoing scientific research and monitoring are focused in certain areas where the actual global pattern of bycatch is no longer representative (Atlantic versus Pacific and Indian oceans). Still, scarce knowledge is available for many species, mainly for those leaving in remote environments. Hence, more information on biology, ecology and population dynamics of chondrichthyes is required to have a better understanding of species-specific bycatch composition occurring in different fishery regions. Despite cartilaginous fish are subjected to international conservation and management measures, only in few countries, advices really take place. More co-operation between fishing nations and effective recommendations is needed on both global and regional scales.

Author details

Sara Bonanomi[1]*, Jure Brčić'[2], Alessandro Colombelli[1], Emilio Notti[1], Jacopo Pulcinella[1] and Antonello Sala[1]

*Address all correspondence to: sara.bonanomi@an.ismar.cnr.it

1 Italian National Research Council (CNR), Institute of Marine Sciences (ISMAR), Ancona, Italy

2 Department of Marine Studies, University of Split, Split, Croatia

References

[1] Stevens JD, Bonfil R, Dulvy NK, Walker PA. The effects of fishing on sharks, rays, and chimaeras (chondrichthyes), and the implications for marine ecosystems. ICES Journal of Marine Science: Journal du Conseil. 2000;**57**:476-494

[2] Cortés E. Incorporating uncertainty into demographic modeling: Application to shark populations and their conservation. Conservation Biology. 2002;**16**:1048-1062

[3] Cortés E, Arocha F, Beerkircher L, Carvalho F, Domingo A, Heupel M, Holtzhausen H, Santos MN, Ribera M, Simpfendorfer C. Ecological risk assessment of pelagic sharks caught in Atlantic pelagic longline fisheries. Aquatic Living Resources. 2010;**23**:25-34

[4] Kieran Kelleher, Food and Agriculture Organization of the United Nations Food & Agriculture Org., Business & Economics, 2005; p. 131

[5] Lawson T. Estimation of catch rates and catches of key shark species in tuna fisheries of the Western and Central Pacific Ocean using observer data. In: Information Paper EB IP–02. Seventh Regular Session of the Scientific Committee of the WCPFC; 9th–17th August 2011; Pohnpei, FSM

[6] Dulvy NK, Fowler SL, Musick JA, Cavanagh RD, Kyne PK, Harrison LR, Carlson JK, Davidson LNK, Fordham SV, Francis MP, Pollock CM, Simpfendorfer CA, Burgess GH, Carpenter KE, Compagno LJV, Ebert DA, Gibson C, Heupel MR, Livingstone SR, Sanciangco JC, Stevens JD, Valenti S, White WT. Extinction risk and conservation of the world's sharks and rays. eLife. 2014;**3**:e00590

[7] Dent F, Clarke S. State of the global market for shark products. FAO Fisheries and Aquaculture Technical Paper; 2015. p. 590

[8] ICCAT. Report of the standing committee on research and statistics (SCRS). International Commission for the Conservation of Atlantic Tunas (ICCAT) SCRS Report; 2009

[9] ICCAT. Report of the Standing Committee on Research and Statistics (SCRS). International Commission for the Conservation of Atlantic Tunas (ICCAT) SCRS Report; 2012

[10] James KC, Lewison RL, Dillingham PW, Curtis KA, Moore JE. Drivers of retention and discards of elasmobranch non-target catch. Environmental Conservation. 2015;**43**:3-12

[11] Oliver S, Braccini M, Newman SJ, Harvey ES. Global patterns in the bycatch of sharks and rays. Marine Policy. 2015;**54**:86-97

[12] Ferretti F, Myers RA, Serena F, Lotze HK. Loss of large predatory sharks from the Mediterranean Sea. Conservation Biology. 2008;**22**:952-964

[13] Ferretti F, Worm B, Britten GL, Heithaus MR, Lotze HK. Patterns and ecosystem consequences of shark declines in the ocean. Ecology Letters. 2010;**13**:1055-1071

[14] Dulvy NK, Metcalfe JD, Glanville J, Pawson MG, Reynolds JD. Fishery stability, local extinctions, and shifts in community structure in skates. Conservation Biology. 2000;**14**: 283-293

[15] Burgess GH, Carvalho JF, Imhoff JL. An evaluation of the status of the largetooth sawfish, *Pristis perotteti*, based on historic and recent distribution and qualitative observations of abundance. Florida Museum of Natural History Report; 2009

[16] Worm B, Davis B, Kettemer L, Ward-Paige CA, Chapman D, Heithaus MR, Kessel ST, Gruber SH. Global catches, exploitation rates, and rebuilding options for sharks. Marine Policy. 2013;**40**:194-204

[17] Bonfil R. Overview of world elasmobranch fisheries. Food and Agriculture Organization of the United Nations, Rome. FAO Fisheries Technical Paper; 1994. p. **341**

[18] Alverson D, Freeberg M, Pope JG, Murawski SA. A global assessment of fisheries bycatch and discards. FAO Fishery Technical Paper. 1994;**339**:1-233

[19] Beerkircher LR, Brown CJ, Lee D. SEFSC pelagic observer program data summary for 1992-2000. NMFS-SEFC-486 NOAA Technical Memorandum; 2002. p. **26**

[20] Lewison RL, Crowder LB, Read AJ, Freeman SA. Understanding impacts of fisheries bycatch on marine megafauna. Trends in Ecology & Evolution. 2004;**19**:598-604

[21] Keene KF. SEFSC Pelagic Observer Program Data Summary for 2007-2011. NOOA Technical Memorandum NMFS-SEFSC; 2016. p. 687

[22] Molina JM, Cooke SJ. Trends in shark bycatch research: Current status and research needs. Reviews in Fish Biology and Fisheries. 2012;**22**:719-737

[23] Campana SE, Joyce W, Manning MJ. Bycatch and discard mortality in commercially caught blue sharks *Prionace glauca* assessed using archival satellite pop-up tags. Marine Ecology Progress Series. 2009;**387**:241-253

[24] Campana SE, Warren J, Malcom PF, Manning MJ. Comparability of blue shark mortality estimates for the Atlantic and Pacific longline fisheries. Marine Ecology Progress Series. 2009;**396**:161-164

[25] Coelho R, Fernandez-Carvalho J, Lino PG, Santos MN. An overview of the hooking mortality of elasmobranchs caught in a swordfish pelagic longline fishery in the Atlantic Ocean. Aquatic Living Resources. 2012;**25**:311-319

[26] Stevens J. *Prionace glauca*. The IUCN Red List of Threatened Species. 2009;e.T39381A10222811

[27] Queiroz N, Humphries NE, Mucientes G, Hammerschlag N, Lima FP, Scales KL, Miller PI, Sousa LL, Seabra R, Sims DW. Ocean-wide tracking of pelagic sharks reveals extent of overlap with longline fishing hotspots. Proceedings of the National Academy of Sciences. 2016;**113**:1582-1587

[28] Clarke SC, Milner-Gulland EJ, Bjørndal T. Perspective: Social, economic and regulatory drivers of the shark fin trade. Marine Resource Economics. 2007;**22**:305-327

[29] Clarke SC, Harley SJ, Hoyle SD, Rice JS. Population trends in Pacific oceanic sharks and the utility of regulations on shark finning. Conservation Biology. 2013;**27**:197-209

[30] Laptikhovsky VV. Survival rates for rays discarded by the bottom trawl squid fishery off the Falkland Islands. Fish Bulletin. 2004;**102**:757-759

[31] Damalas D, Vassilopoulou V. Chondrichthyan by-catch and discards in the demersal trawl fishery of the central Aegean Sea (Eastern Mediterranean). Fisheries Research. 2011;**108**:142-152

[32] Enever R, Revill AS, Caslake R, Grant A. Discard mitigation increases skate survival in the Bristol Channel. Fisheries Research. 2010;**102**:9-15

[33] Clarke SC, McAllister MK, Milner-Gulland EJ, Kirkwood GP, Michielsens CGJ, Agnew DJ, Pikitch EK, Nakano H, Shivji MS. Global estimates of shark catches using trade records from commercial markets. Ecology Letters. 2006;**9**:1115-1126

[34] ICES. Report of the Workshop to compile and refine catch and landings of elasmobranchs (WKSHARKS); 19-22 January 2016; Lisbon, Portugal. ICES CM 2016/ACOM: **40**. 69 pp

[35] Hueter RE, Manire CA, Tyminski JP, Hoenig JM, Hepworth DA. Assessing mortality of released or discarded fish using a logistic model of relative survival derived from tagging data. Transactions of the American Fisheries Society. 2006;**135**:500-508

[36] ICES. ICES implementation of Advice for Data-limited stocks in 2012 in its 2012 Advice. ICES CM 2012/ACOM: **68**. 42 pp

[37] Ellis J. *Raja clavata*. The IUCN Red List of Threatened Species. 2016; e.T39399A103110667

[38] Walker TI, Gason AS. SESSF monitoring data management, reporting and documentation 2006/07. Final report to Australian Fisheries Management Authority. Project No. R2006/812. Victoria, Australia: Primary Industries Research Victoria, Department of Primary Industries, Queenscliff; 2009. pp. vii + 177

[39] Ebert DA, Stehmann MFW. Sharks, Batoids, and Chimaeras of the North Atlantic. Food and Agriculture Organization of the United Nations; 2013

[40] Calis E, Jackson EH, Nolan CP, Jeal F. Preliminary age and growth estimates of the Rabbitfish, Chimaera monstrosa, with implications for future resource management. Journal of North West Atlantic Fisheries Science. 2005;**35**:21

[41] Cavanagh RD, Gibson C. Overview of the Conservation Status of Cartilaginous Fishes (Chondrichthyans) in the Mediterranean Sea. Gland, Switzerland and Malaga, Spain: IUCN; 2007. p. vi + 42

[42] Dagit DD, Hareide N, Clò S. *Chimaera monstrosa*. The IUCN Red List of Threatened Species. 2007;e.T63114A12610445

[43] Frisch AJ, Ireland M, Rizzari JR, Lönnstedt OM, Magnenat KA, Mirbach CE, Hobbs JPA. Reassessing the trophic role of reef sharks as apex predators on coral reefs. Coral Reefs. 2016;**35**:459

[44] Bornatowski H, Navia AF, Braga RN, Abilhoa V, Corrêa MFM. Ecological importance of sharks and rays in a structural foodweb analysis in Southern Brazil. ICES Journal of Marine Science. 2014;**71**:1586-1592

[45] Simpfendorfer CA, Kyne PM. Limited potential to recover from overfishing raises concerns for deep-sea sharks, rays and chimaeras. Environmental Conservation. 2009;**36**: 97-103

[46] Kyne PM, Simpfendorfer CA. Deepwater Chondrichthyans. Sharks and their Relatives. II. Biodiversity, Adaptative Physiology, and Conservation. New York: CRC Press; 2010. pp. 37-114

[47] Baum JK, Worm B. Cascading top-down effects of changing oceanic predator abundances. Journal of Animal Ecology. 2009;**78**:699-714

[48] Gallagher AJ, Orbesen ES, Hammerschlag N, Serafy JE. Vulnerability of oceanic sharks as pelagic longline bycatch. Global Ecology and Conservation. 2014;**1**:50-59. ISSN 2351-9894. http://dx.doi.org/10.1016/j.gecco.2014.06.003

[49] Ward P, Myers R. Inferring the depth distribution of catchability for pelagic longlines correcting for variations in the depth of longline fishing. Canadian Journal of Fisheries and Aquatic Sciences. 2005;**62**:1130-1142

[50] Shepherd TD, Myers RA. Direct and indirect fishery effects on small coastal elasmobranchs in the Northern Gulf of Mexico. Ecology Letters. 2005;**8**:1095-1104

[51] Myers RA, Baum JK, Shepherd TD, Powers SP, Peterson CH. Cascading effects of the loss of apex predatory sharks from a coastal ocean. Science. 2007;**315**:1846-1850

[52] ICCAT. Compendium Management Recommendations and Resolutions Adopted by Iccat for the Conservation of Atlantic Tunas and Tunalike Species. International Commission for the Conservation of Atlantic Tunas (ICCAT); 2014

[53] ICCAT. Recommendation by Iccat on Improvement of Compliance Review of Conservation and Management Measures Regarding Sharks Caught in Association with Iccat Fisheries. International Commission for the Conservation of Atlantic Tunas (ICCAT); 2016

[54] Campana SE, Joyce W, Fowler M, Showell M. Discards, hooking, and post-release mortality of porbeagle (*Lamna nasus*), shortfin mako (*Isurus oxyrinchus*), and blue shark (*Prionace glauca*) in the Canadian pelagic longline fishery. ICES Journal of Marine Science. 2015;**73**:520-528

[55] Smith SE, Au DW, Show C. Intrinsic rebound potentials of 26 species of Pacific sharks. Marine and Freshwater Research. 1998;**49**:663-678

[56] Frisk MG, Miller TJ, Fogarty MJ. Estimation and analysis of biological parameters in elasmobranch fishes: A comparative life history study. Canadian Journal of Fisheries and Aquatic Sciences. 2001;**58**:969-981

[57] Lucchetti A, Sala A, Kholeif SEA, Notti E. Towards Sustainable Fisheries Management: A Perspective of Fishing Technology Weaknesses and Opportunities with a Focus on the Mediterranean Fisheries. New York: Nova Publisher Book; 2015. ISBN: 978-1-63463-816-6

[58] Favaro B, Côté IM. Do by-catch reduction devices in longline fisheries reduce capture of sharks and rays? A global meta-analysis. Fish and Fisheries. 2015;**16**:300-309

[59] Dai XJ, Zheng Y, Jiang RL, Xu LX. Shark by-catch observation in ICCAT waters by the Chinese longline observers in 2007. Collection volume scientific papers. ICCAT. 2009;**64**: 1741-1745

[60] Brčić J, Herrmann B, De Carlo F, Sala A. Selective characteristics of a shark-excluding grid device in a Mediterranean trawl. Fisheries Research. 2016;**172**:352-360

[61] Ferretti F, Osio GC, Jenkins CJ, Rosenberg AA, Lotze HK. Long-term change in a mesopredator community in response to prolonged and heterogeneous human impact. Scientific Reports. 2013;**3**:1057

[62] Campana SE, Ferretti F, Rosenberg A. Sharks and other elasmobranchs. The first global integrated marine assessment, World Ocean Assessment I, United Nations; 2016; 1437-1451

[63] Amande MJ,Chassot E, Chavance P, Pianet R. Silky Shark (*Carcharhinus falciformis*) by Catch in the French Tuna Purse-seine Fishery of the Indian Ocean. Mahe, Seychelles: IOTCWPEB008/016 Indian Ocean Tuna Commission; 2008. p. 22

[64] Western Pacific Tuna Commission (WCPFC). Tuna Fishery Yearbook. 2011. Availablefrom: http://www.wcpfc.int/doc/2011/wcpfc-tuna-fishery-yearbook-2011-ex cel-files-170kb [Accessed: 10.04.13]

[65] Sala A, Lucchetti A, Perdichizzi A, Herrmann B, Rinelli P. Is square-mesh better selective than larger mesh? A perspective on the management for Mediterranean trawl fisheries. Fisheries Research. 2015;**161**:182-190

[66] Sala A, Lucchetti A, Affronte M. Effects of turtle excluder devices on bycatch and discards reduction in the demersal fisheries of Mediterranean Sea. Aquatic Living Resources. 2011;**24**:183-192

[67] Lucchetti A, Sala A. An overview of loggerhead sea turtle (*Caretta caretta*) bycatch and technical mitigation measures in the Mediterranean Sea. Reviews of Fish Biology and Fisheries. 2010;**20**:141-161

[68] Techera EJ, Klein N. Fragmented governance: Reconciling legal strategies for shark conservation and management. Marine Policy. 2011;**35**:73-78

[69] Fischer J, Erikstein K, D'Offay B, Barone M, Guggisberg S. Review of the Implementation of the International Plan of Action for the Conservation and Management of Sharks. FAO Fisheries and Aquaculture Circular No. 1076. Rome: FAO; 2012. p. 120

[70] FAO. Food and Agricultural Organization. International Plan of Action for the conservation and management of sharks (IPOA). 1999

[71] Watson JW, Epperly SP, Shah AK, Foster DG. Fishing methods to reduce sea turtle mortality associated with pelagic longlines. Canadian Journal of Fisheries and Aquatic Sciences. 2005;**981**:965-981

[72] Kim SS, Moon DY, An DH, Koh JR. Comparison of circle hook and J hook catch rate for target and bycatch species taken in the Korean tuna longline fishery. In: Paper presented at the Second Regular Session of the Scientific Committee; August 7-18, 2007; Manila, Philippines. Western and central Pacific Fisheries commission WcPFcSc2-eB WP-112

[73] Gilman E, Kobayashi D, Swenarton T, Brothers N, Dalzell P, Kinan-Kelly I. Reducing sea turtle interactions in the Hawaii based longline swordfish fishery. Biological Conservation. 2007;**139**:19-28

[74] Promjinda S, Siriraksophon S, Darumas N, Chaidee P. Efficiency of the circle hook in comparison with J-hook in longline fishery. SEAFDEC—The Ecosystem Based Fishery Management in the Bay of Bengal. 2008.

[75] Carruthers E, David H, Schneider C, Neilson JD. Estimating the odds of survival and identifying mitigation opportunities for common bycatch in pelagic longline fisheries. Biological Conservation. 2009;**142**:2620-2630. DOI: 10.1016/j.biocon.2009.06.010

[76] Alfonso A, Hazin F, Carvalho F, Pacheco J, Hazin H, Kerstetter D, Murie D, Burgess G. Fishing gear modifications to reduce elasmobranch mortality in pelagic and bottom longline fisheries off North East Brazil. Fisheries Research. 2011;**108**:336-343

[77] Curran D, Bigelow K. effects of circle hooks on pelagic catches in the Hawaii-based tuna longline fishery. Fisheries Research. 2011;**109**:265-275

[78] Gilman E, Huang HW. Review of effects of pelagic longline hook and bait type on sea turtle catch rate, anatomical hooking position and at-vessel mortality rate. Reviews in Fish Biology and Fisheries. 2017;**27**. 1-10. DOI: 10.1007/s11160-016-9447-9

[79] Petersen SL, Honig MB, Ryan PG, Underhill LG, Compagno LJV. Pelagic shark bycatch in the tuna- and swordfish directed longline fishery off Southern Africa. African Journal of Marine Science. 2009;**31**:215-225

[80] Galeana-Villasenor I, Galvan-Magana F, Santana-Hernandez H. Fishing by hooks in longliners from the Mexican Pacific Ocean: Effects in the catch rate and weight of sharks and other species. Revista de Biologia Marina y Oceanografia. 2009;**44**:163-172

[81] Branstetter S, Musick JA. Comparisons of shark catch rates on longlines using rope/steel (Yankee) and monofilament gangions. Marine Fisheries Review. 1991;**55**:4-9

[82] Stone HH, Dixon LX. a comparison of catches of swordfish, *Xiphias gladius*, and other pelagic species from Canadian longline gear configured with alternating monofilament and multifilament nylon gangions. Oceans. 2001;**216**:210-216

[83] Ward P, Lawrence E, Darbyshire R, Hindmarsh S. Largescale experiment shows that nylon leaders reduce shark bycatch and benefit pelagic longline fishers. Fisheries Research. 2008;**90**:100-108

[84] Vega R, Licandeo R. The effect of American and Spanish longline systems on target and non-target species in the eastern South Pacific swordfish fishery. Fisheries Research. 2009;**98**:22-32

[85] Williams PG. Shark and related species catch in tuna fisheries of the tropical Western and central Pacific Ocean. In: Shotton R, editor. Case Studies on the Management of Elasmobranch Fisheries. Rome: FAO Technical Paper; 1999. p. 378. Part 1

[86] Hinke JT, Kaplan IC, Aydin K, Watters GM, Olson R, Kitchell JF. Visualizing the food web effects of fishing for tunas in the Pacific Ocean. Ecology and Society. 2004;**9**:1

[87] Gilman EL, Clarke S, Brothers N, Alfaro-Shigueto J, Mandelman J, Mangel J, Petersen S, Piovano S, Thomson N, Dalzell P, Donoso M, Goren M, Werner T. Shark Depredation and Unwanted Bycatch in Pelagic Longline Fisheries: Industry Practices and Attitudes, and Shark Avoidance Strategies. Honolulu: Western Pacific Regional Fishery Management Council; 2007

[88] Carrier, J.C.; Musick, J.A.; Heithaus, M.R. (Ed.) (2004). Biology of sharks and their relatives. CRC Marine Biology Series. CRC Press: Boca Raton. ISBN 0-8493-1514-X. 596 pp

[89] Kaimmer S, Stoner A. Field investigation of rare-earth metal as a deterrent to spiny dogfish in the Pacific halibut fishery. Fisheries Research. 2008;**94**:43-47

[90] O'Connell CP, Stroud EM, He P. The emerging field of electrosensory and semiochemical shark repellents: Mechanisms of detection, overview of past studies, and future directions. Ocean & Coastal Management. 2014;**97**:2-11

[91] O'Connell CP, Andreotti S, Rutzen M, Meÿer M, He P. The use of permanent magnets to reduce elasmobranch encounter with a simulated beach net. 2. The great white shark (*Carcharodon carcharias*). Ocean & Coastal Management. 2014;**97**:20-28

[92] Hoey JJ, Moore NE. Multi-species catch characteristics for the U.S. Atlantic pelagic longline fishery: Captain's report. National Marine Fisheries-NOAA-NMFS. Marfin Grant-NA77FF0543, (SK) Grant NA86FD0113. 1999

[93] Coelho R, Erzini K, Bentes L, Correia C, Lino PG, Monteiro P, Ribeiro, Gonçalves JMS. Semi-pelagic longline and trammel net elasmobranch catches in Southern Portugal: Catch composition, catch rates, and discards. Journal of North West Atlantic Fishery Science. 2005;**37**:531-537

[94] Carlson J, Cortes E. Gillnet selectivity of small coastal sharks off the South-Eastern United States. Fisheries research. 2003;**60**:405-414

[95] McAuley RB, Simpfendorfer CA, Wright IW. Gillnet Mesh Selectivity of the sandbar shark (*Carcharhinus plumbeus*): Implications for fisheries management. ICES Journal of Marine Science. 2007;**64**:1702-1709

[96] Thorpe T, Frierson D. Bycatch mitigation assessment for sharks caught in coastal anchored gillnets. Fisheries Research. 2009;**98**:102-112

[97] Hall SJ, Mainprize BM. Managing by-catch and discards: How much progress are we making and how can we do better? Fish and Fisheries. 2005;**6**:134-155

[98] Brewer DD, Heales D, Milton D, Dell Q, Fry G, Venables B, Jones P. The impact of turtle excluder devices and bycatch reduction devices on diverse tropical marine communities in Australia's Northern prawn trawl fishery. Fisheries Research. 2006;**81**:176-188

[99] Zeeberg J, Corten AD, Graaf ED. Bycatch and release of pelagic megafauna in industrial trawler fisheries off NorthWest Africa. Fisheries Research. 2006;**78**:186-195

[100] Baker B, Hamilton S, McIntosh R, Finley L. Technical Review: Development and Application of Bycatch Mitigation Devices for Marine Mammals in Mid-Water Trawl Gear. Report prepared for the Department of the Environment (on behalf of the expert panel); 2014. p. 12

[101] Franco J, Moreno G, Lopez J, Sancristobal I. Testing new designs of drifting fish aggregating device (DFaD) in Eastern Atlantic to reduce turtle and shark mortality. International Commission for the Conservation of Tunas SCRS/2011/067

[102] Schaefer KM, Fuller DW. An overview of the 2011 iSSF/iaTTc research cruise for investigating potential solutions for reducing fishing mortality of undesirable sizes of bigeye and yellowfin tunas and sharks in purse-seine sets on drifting FADs. In: Scientific Committee Seventh Regular Session; August 9 17, 2011; Federated States of Micronesia. WcPFc-Sc7-2011/eB-WP-13. p. 5

[103] Kondel J, Rusin J. Report of the second workshop on bycatch reduction in the ETP purse-seine fishery. NMFS Administrative Report LJ-07-04. 2007

[104] Dagorn L. Mitigating bycatch of sharks and finfish by tropical tuna purse seiners using FADs. ISSF Workshop on Bycatch; Brisbane, Australia; June 26, 2010

[105] ICCAT. ICCAT Recommendations on Closed Area/Season for Fishing with FADs in Eastern Tropical Atlantic (9801) and on Closed Area/Season to FADs (99-1). Madrid: ICCAT Secretariat; 1999

[106] STECF. STECF Report of the SGMOS-09-05 Working Group on Fishing Effort Regimes Regarding Annex IIA of TAC & Quota Regulations and Celtic Sea. 2009.

[107] Mandelman JW, Farrington MA. The estimated short-term discard mortality of a trawled elasmobranch, the spiny dogfish (*Squalus acanthias*). Fisheries Research. 2007;**83**: 238-245

[108] Rulifson RA. Spiny dogfish mortality induced by gill-net and trawl capture and tag and release. North American Journal of Fisheries Management. 2007;**27**:279-285

[109] Hareide NR, Garnes G. The distribution and catch rates of deep-water fish along the Mid-Atlantic Ridge from 43 to 61 N. Fisheries Research. 2001;**519**:297-310

[110] Fernandes AC, Prista N, Jardim E, Silva D, Ferreira AP. Results from the 2010 Portuguese onboard sampling programme of the deepwater longline fleet with an emphasis on elasmobranch species. Working Document presented in the ICES Working Group on Elasmobranch Fishes (WGEF); 2011

[111] Lagarto N, Moura T, Farias I, Figueiredo I. Sampling information from deep-water sharks caught in the black scabbardfish fishery from mainland Portugal. Working Group on Elasmobranch Fishes (WGEF); 2012

[112] Nieto A, Ralph GM, Comeros-Raynal MT, Kemp J, García Criado M, Allen DJ, Dulvy NK, Walls RHL, Russell B, Pollard D, García S, Craig M, Collette BB, Pollom R, Biscoito M, Labbish Chao N, Abella A, Afonso P, Álvarez H, Carpenter KE, Clò S, Cook R, Costa MJ, Delgado J, Dureuil M, Ellis JR, Farrell ED, Fernandes P, Florin A-B, Fordham S, Fowler S, Gil de Sola L, Gil Herrera J, Goodpaster A, Harvey M, Heessen H, Herler J, Jung A, Karmovskaya E, Keskin C, Knudsen SW, Kobyliansky S, Kovačić M, Lawson JM, Lorance P, McCully Phillips S, Munroe T, Nedreaas K, Nielsen J, Papaconstantinou C, Polidoro B, Pollock CM, Rijnsdorp AD, Sayer C, Scott J, Serena F, Smith-Vaniz WF, Soldo A, Stump E, Williams JT. European Red List of Marine Fishes. Luxembourg: Publications Office of the European Union; 2015

[113] Dolgov AV, Drevetnyak KV, Gusev EV. The status of skate stocks in the Barents Sea. Journal of Northwest Atlantic Fishery Science. 2005;**35**:1-13

[114] Albert OT, Vollen T. A major nursery area around the Svalbard archipelago provides recruits for the stocks in both Greenland halibut management areas in the Northeast Atlantic. ICES Journal of Marine Science: Journal du Conseil. 2014;**3**:fsu191

[115] Hoey JJ, Pritchard E, Brown C, Showell M. Pelagic shark abundance indices based on fishery-dependent and fishery-independent data from the Western North Atlantic. Collected Volumes Scientific Papers. ICCAT. 2002;**54**:1199-1211

[116] Campana SE, Brading J, Joyce W. Estimation of pelagic shark bycatch and associated mortality in Canadian Atlantic fisheries. DFO Canadian Science Advisory Secretariat; 2011. p. 67

[117] Campana SE. Transboundary movements, unmonitored fishing mortality, and ineffective international fisheries management pose risks for pelagic sharks in the North West Atlantic. Canadian Journal of Fisheries and Aquatic Sciences. 2016;**73**:1-9

[118] Herndon A, Gallucci VF, DeMaster D, Burke W. The case for an international commission for the conservation and management of sharks (ICCMS). Marine Policy. 2010;**34**:1239-1248

[119] Cortés E, Domingo A, Miller P, Forselledo R, Mas F, Arocha F, Campana S, Coelho R, Da Silva R, Hazin FHV, Holtzhausen H, Keene K, Lucena F, Ramirez K, Santos MN, Semba-Murkami Y, Yokawa K. Expanded ecological risk assessment of pelagic sharks caught in Atlantic pelagic longline fisheries. Collective Volume of Scientific Papers ICCAT. 2015;**71**:2637-2688

[120] Poisson F, Crespo FA, Ellis JR, Chavance P, Pascal B, Santos MN, Séret B, Korta M, Coelho R, Ariz J, Murua H. Technical mitigation measures for sharks and rays in fisheries for tuna and tuna-like species: Turning possibility into reality. Aquatic Living Resources. 2016;**29**:402

[121] Castro JI, Woodly CM, Bredek RL. A preliminary evaluation of the status of shark species. FAO Fisheries Technical Paper, 380. Rome, Italy: FAO; 1999

[122] Dell'Apa A, Kimmel DG, Clò S. Trends of fish and elasmobranch landings in Italy: Associated management implications. ICES Journal of Marine Science. 2012;**69**: 1045-1052

[123] Sala A, Brčić J, Conides A, De Carlo F, Klaoudatos D, Grech D, Lucchetti A, Mayans A, Notti E, Paci N, Salom S, Sartor P, Sbrana M, Soler I, Spedicato MT, Virgili M. Technical specifications of Mediterranean trawl gears (myGears). Final project report, financed by the European Commission through the Framework service contract for Scientific Advice and other services for the implementation of the Common Fisheries Policy in the Mediterranean (Contract MARE/2009/05-Lot 1). 2013. p. 519

[124] Guijarro B, Massutì E. Selectivity of diamond- and square-mesh codends in the deepwater crustacean trawl fishery off the Balearic Islands (Western Mediterranean). ICES Journal of Marine Science. 2006;**63**:52-67

[125] Ordines F, Massutí E, Guijarro B, Mas R. Diamond vs. square mesh codend in a multispecies trawl fishery of the Western Mediterranean: Effects on catch composition, yield, size selectivity and discards. Aquatic Living Resources. 2006;**19**:329-338

[126] Ferretti F, Myers RA. By-Catch of sharks in the Mediterranean Sea: Available mitigation tools. In: Proceedings of the Workshop on the Mediterranean Cartilaginous Fish with emphasis on Southern and Eastern Mediterranean. Istanbul, Turkey. 2006. pp. 158-169

[127] UNEP-MAP-RAC/SPA. Status and Conservation of Fisheries in the Adriatic Sea. By H. Farrugio and Alen Soldo. Draft internal report for the purposes of the Mediterranean Regional Workshop to Facilitate the Description of Ecologically or Biologically Significant Marine Areas, Malaga, Spain, 7 11 April 2014.

[128] Fortuna CM, Vallini C, Filidei E, Ruffino M, Consalvo I, Di Muccio S, Gion C, Scacco U, Tarulli E, Giovanardi O, Mazzola A. By-catch of cetaceans and other species of conservation concern during pair trawl fishing operations in the Adriatic Sea (Italy). Chemistry and Ecology. 2010;**26**:65-76

[129] Wintner S. Aetomylaeus bovinus. The IUCN Red List of Threatened Species. 2016; e.T60127A104022824. Available form: http://dx.doi.org/10.2305/IUCN.UK.2016

[130] Zogaris S, Dussling U. On the occurrence of the Bull Ray *Pteromylaeus bovinus* (Chondrichthyes: Myliobatidae) in the Amvrakikos Gulf, Greece. Mediterranean Marine Science. 2010;**11**:177-184

[131] Aldebert Y. Demersal resources of the Gulf of Lions (NW Mediterranean). Impact of exploitation on fish diversity. Vie Milieu. 1997;**47**:275-284

[132] Di Natale A. By-catch of shark species in surface gear used by the Italian fleet for large pelagic species. Collective Volume of Scientific Papers ICCAT. 1998;**48**:138-140

[133] Megalofonou P, Damalas D, Yannopoulos C. Composition and abundance of pelagic shark by-catch in the Eastern Mediterranean Sea. Cybium. 2005;**29**:135-140

[134] Stergiou KI, Moutopoulos DK, Erzini K. Gill net and longlines fisheries in Cyclades waters (Aegean Sea): Species composition and gear competition. Fisheries Research. 2002;**57**:25-37

[135] Serena F. Field Identification Guide to the Sharks and Rays of the Mediterranean and Black Sea. FAO Species Identification Guide for Fishery Purposes. Rome: FAO; 2005; p. 97

[136] Bradai MN, Saidi B, Enajjar S. Elasmobranchs of the Mediterranean and Black sea: Status, ecology and biology. Bibliographic analysis. Studies and Reviews. General Fisheries Commission for the Mediterranean. No. 91. Rome: FAO; 2012. p. 103

Chapter 4

The Importance of Academic Research in the Field of Shark-Human Interactions: A Three-Pronged Approach to a Better Understanding of Shark Encounters

Erich Ritter and Raid Amin

Additional information is available at the end of the chapter

http://dx.doi.org/10.5772/intechopen.69409

Abstract

One of the least understood areas of shark research is the interaction between animals and humans, from factors influencing face-to-face encounters to the causes of incidents and behavior patterns. Although some of these questions can be addressed through studies outside of the water, most interaction-related questions have to be tested by direct observation of the sharks. The main goal of our research is to better understand sharks, and what influences them the most when they are in the vicinity of a human being. Our improved understanding of this animal substantiates the notion that sharks are no different from other animal species and do not reflect the still too often presented media monster. Understanding the true nature of this animal transforms this wrongfully described creature into the rather shy and intelligent animal that it truly is.

Keywords: behavior, body language, shark, shark bites, shark-human interaction

1. Introduction

The possibility of the presence of a shark can easily enter a person's mind when entering the sea for recreational activities, such as swimming and surfing [1]. For most people, the thought of encountering a shark is frightening and has become even more so due to sensational media coverage, and since we were shown very clearly in the movie JAWS how things could turn out when a person meets a hungry shark. But despite the impact of gory movies, media exaggeration and the fact that most people understand that such encounters are different from movies like JAWS, rational thinking still ceases to exist in the presence of a shark and a single thought occupies the mind; how to get away from a shark as quickly as possible and get out of the water.

Despite the fear of a potential shark encounter when people decide to enter the water, the proper response of how to withdraw effectively in such a scenario is not well known [2]. Unfortunately, the plethora of unsubstantiated suggestions flooding the literature [3–5] and Internet make the knowledge of how to correctly interact with a shark and identification of the correct, and more importantly, well-tested ways of leaving the water without getting harmed more than just the proverbial needle in the haystack.

These types of situations are addressed in the field of experimental shark-human interactions. Whenever an interaction between a human being and a shark occurs, knowingly or unknowingly, from a person's viewpoint, this type of research explores the core of this type of event. Shark-human interaction studies address various questions, from the validity of the "mistaken identity theory" [6, 7] to the interpretation of wound pictures [8, 9], and approach pattern and body language of sharks in the vicinity of humans [10, 11]. Despite the broad spectrum of questions in this research field, it mostly boils down to what the intention of these animals are while in the vicinity of a human being. Although we will never be able to know for sure what a shark truly thinks, experiments and analytical tools can be used to get as close as possible to the true intention of a shark during such circumstances [12].

The following sections reflect a summary of our research within the shark-human interaction field. Behavior identifications and forensic analyses as well as the various aspects dealing with the human mindset will highlight the most important findings and show how these findings can improve the way a person thinks and acts when it comes to interactions with sharks.

1.1. JAWS is not out there

For decades, sharks have been falsely portrayed by the media more than any other animal [13–15]. The movie JAWS, which was released in 1975, has been by far the biggest influence on public perception of sharks. One of the main reasons why this movie was seen as representative of the nature of sharks was that scientists in that era were not able to counter this general opinion with factual evidence. At the time, very few scientists were studying sharks, and nobody specifically worked in the field of shark-human interactions [16]. Due to these shortcomings, the media had a free rein in their representation of these animals. Granted, today's mainstream media has taken some steps to correct this wrongful image, roughly half of all stories in the media still portray sharks negatively, thus making it impossible to reverse the falsely negative image of these animals. For example, a reflection of this perception is the nearly nonrestricted industrial fishing that led to dwindling shark populations. Sharks are still mostly seen as dangerous and so protective measures are either up to this time nonexistent or not effective enough [17–19]. People tend to follow the old adage that they do not need to protect what they are afraid of. Indeed, ignorance and arrogance have led to the elimination of other animal groups in the past [13, 20], and sharks are in similar danger.

Probably, the most important tool in how to change the attitude of the public is not just education but should specifically address our needs of sharks, and the ticking ecological time bomb which needs to be defused now [21, 22]. It is also to show that we do not have to be afraid of sharks, despite what movies, the media and the lack of scientific knowledge in the 1970s and 1980s made us believe [16]. Understanding animals means spending time observing them closely and trying to comprehend the flow of information between humans and sharks, which can only occur when studying them in their realm. One of the most prominent aspects of any interaction with a shark is the unknown factor of what a shark "does" in a person's vicinity and the reasoning behind these actions [10–12]. Understanding these interactions and the factors influencing these situations are paramount during every activity. However, these scenarios are not frequent, and it remains challenging to collect meaningful research data by just interacting with these animals without setting up some kind of experimental protocol. Thus, it is crucial to create scenarios in the shark's habitat that allow proper scientific evaluation [10–12].

It is not just an understanding of key factors that impact these interactions that is needed, but it is also necessary to examine the variety of nonhuman-oriented behavioral patterns exhibited by sharks, in order to achieve an accurate description. Behaviors seen in sharks such as gaping and chafing are triggered by a variety of factors [23–25] but are not caused by humans, even if they may be close by.

Besides setting well-defined parameters of data collection methods during testable interactions, it is also necessary for the observer to have a detached mental state, as long as it is not at the core of the experiment. Fear affects people on different levels [26, 27], but when it comes to sharks, fear needs to be addressed more than ever since it likely affects the very scenario that is being investigated [28–30]. Although fear most often ceases to exist, once people get used to sharks [2], it can remain present on different levels. Although most of a person's basic fears are projected onto the shark, it is mostly the fear of not knowing how the situation with a shark will develop, and which factors influence the situation, as well as what has to be done, if necessary, that dominates and controls the scenario [2].

In the following sections, we will briefly describe the different behavioral and approach patterns of sharks close to humans, how factors underlying them will influence the scenarios and give an overview of the main behaviors we found. Furthermore, we provide an outlook of how forensic bite analysis and how bite rates along shorelines can be used to complete the picture of the shark that we should spread. Shark-human interactions consist of many facets, and even though we have only just begun to scratch the surface, the emerging picture already reveals a much different image than is so often presented in the mainstream media.

1.2. What is it with surfers and sharks?

Surfing and sharks is a commonly discussed topic along many of the world's beaches with the general assumption that sharks, especially white sharks, mistake surfers for seals being the main cause of incidents [31, 32]. It is true that surfers are the most exposed group when it comes to contact with sharks [33], but the idea of a mistake, as a cause for bites, is not true [6, 7]. We have shown that there is no mistake when a white shark bites a surfer [7]. It mostly happens because the shark does not know what a surfer is, and therefore exploration is the main cause for these bites. Indeed, exploration is also the basis of many other interactions between humans and sharks that end in a conflict [8, 9].

In order to establish exploration as the main cause for white sharks biting surfers, we needed to compare board damages and wounds on surfers with the wounds on pinnipeds, especially seals, which are the white shark's usual prey [34–36]. It was not just the actual outcomes that were important, but we also had to understand the white shark's mindset as it grows up and is confronted with pinnipeds. Pinnipeds are highly agile mammals and demand tremendous hunting skills if a shark is to catch them [37–39]. Although adult white sharks are very effective hunters, when they are small or young, they are far from capable of catching these pinnipeds [31]. Thus, in order to be successful, white sharks need to reach a rather large size, at least 3.5 m, before they are effective at catching this prey [36]. Even then, their success rate remains rather low.

If the assumed "mistaken identity theory" was true, the minimum size of white sharks targeting surfers would be similar to sharks that hunt pinnipeds. However, our research showed that the sharks involved in bites on surfers are often much smaller than this, even less than 2.5 m in length. To reconstruct a white shark's size, we needed appropriate references from the wounds or board damages, preferably adjacent tooth imprints. Since there is a strong correlation between a shark's length and its interdental distances, the measurements taken from surfer wounds or damaged boards allowed us to accurately estimate the size of the white shark involved [40, 41].

Both the size of the white sharks involved in surfer incidents and the severity of the damage done to the body or board were important factors. Bites on pinnipeds are executed with tremendous force by a white shark to prevent the escape of the targeted prey. Therefore, if attacks on surfers were due to a case of mistaken identity, the wounds on surfers and/or board damages would also be expected to be extensive. To investigate this, we created a severity index that allowed the degree of the wound and board damage to be assessed. The scale ranged from 0 to 5, with 0 being the most superficial and 5 representing the most extensive damage to a person or board. If a mistake was the cause of an incident, it could be assumed that the damage would be very high, at least level 4. Nevertheless, we obtained an average score of 1.8, indicating a quite superficial level of damage, certainly insufficient to prevent an escape should the surfers indeed have been pinnipeds. Both of our results, smaller sharks initiating a bite and the rather superficial damage caused, indicate that these incidents are not the result of a mistake, but that the shark purposely initiated a bite with the likely motivation of getting more information of what a surfer might be. Two-thirds of all bites were superficial in nature, and the remaining one-third either were more severe or showed multiple bites. Here as well, a mix-up can be excluded, even for the more intense outcomes. For an explanation, motivating factors such as target practice or object play needed to be considered. Both of these theories require further examination but are likely causes for the more severe bites.

Although "mistaken identity theory" can be ruled out as a potential cause for surfer bites by white sharks, there are other shark species that are involved with surfers, for example on the East coast of Florida where there are blacktips (*Carcharhinus limbatus*), silkies (*Carcharhinus falciformis*), spinner sharks (*Carcharhinus brevipinna*) and possibly even bull sharks (*Carcharhinus leucas*) [42]. The carcharhinid sharks' diet primarily consists of fish but not pinnipeds, and they do not even live near pinnipeds, thus the "mistaken identity theory" in shark incidents involving surfers along the Florida coast could not even be considered.

http://dx.doi.org/10.5772/intechopen.69409

The "mistaken identity theory" arose at a time when sharks were still thought of as having rather low intelligence and being primarily driven by their instincts. From that viewpoint, a mistake could make sense but even so, considering the rather long coevolution between the white sharks and pinnipeds, it is difficult to accept that these sharks could be misled, since a surfer looks rather dissimilar to the outline of a pinniped when seen from below. But even if any silhouette resemblance could be established, a shark would have to be pretty much right below the surfer when looking up, because the shallower the approach angle the less likely the silhouette will fit the outline of a pinniped. It is a fact that most surfers are indeed bitten in rather shallow water that would not even allow a shark to swim straight up from directly below once it notices the surfer.

The phenomenon of sharks biting surfers needs further examination, not just from a species viewpoint, like the carcharhinid sharks, but also from a viewpoint of the approach angle. As we will see in the next section, sharks are able to understand human body orientation [10, 11], which leads to an interesting question; can white sharks and carcharhinid sharks determine the orientation of surfers when sitting or lying on their boards?

1.3. Understanding human body orientation

Whenever people recall an incident with a shark or just a sighting, they use their own body orientation as a reference for the shark's approach direction, i.e., the shark came from behind [43–45]. Although it is commonly accepted that a person's body orientation could be used to explain a shark's approach direction, whether sharks are indeed able to understand such orientation has never been questioned [10, 12].

People have been intensively using the ocean's shorelines for fishing and harvesting purposes for only a few hundred years, making it very unlikely for even shore-oriented shark species to have learned human body orientation. Not having coevolved seems to make it rather unlikely that sharks would be able to understand a person's body direction. However, our research showed that this is not true. Sharks, in this case Caribbean reef sharks, *Carcharhinus perezi*, are able to do just that [10], and it is highly likely that direction of an approaching shark is thus influenced by the person's body orientation [11]. We were also able to show that when changing our body position from horizontal to vertical, e.g., standing up in a shallow area, that bull sharks stayed significantly farther away than when the test subject was lying flat at the same spot. This raises the question of what these sharks focus on to determine human body orientation, and what causes a reaction in them when changing a human's body position. It is known that different species of mammals can identify human body and face orientation [46–48] and are even able to follow a person's gaze and eye movements [47, 49]. It needs to be determined whether sharks are also able to make such detailed observations. However, considering that a swim pattern determination of a shark in the vicinity of a person can take place beyond the close visual range, other factors are likely to come into play. Although there remains much to be understood when it comes to shark-human interactions, the shark's vision must play a crucial role. We also observed that once eye contact was interrupted during an encounter, e.g., by lifting a camera in front of the person's face, that sharks veered away. Whatever it is, sharks seem to react to human eye contact, but there must be something else when the shark is close to the visual range. Of course, one could argue that human locomotion or the direction a person is moving is the trigger of such decision-making and has nothing to with a person's visual system. Should the visibility be low, a shark could indeed follow whirls generated in the water by a person's feet or hands. Thus, the recognition of the pure motion direction of a person seems more likely to give a shark a sense of a person's orientation in the water. But such an idea falls short when a person is not moving.

When sharks approach humans, they prefer to come up from behind [10]. But not just that, the larger the animals, the closer they seem to swim to the sea floor [12] should the person be close to it. We investigated the possibility that this may reflect some kind of stealth behavior. However, biting the person at the end of such an approach does not seem to be the purpose of stealth in this situation but is rather used so that the shark remains unseen, thus free to observe and learn in safety.

Results from our studies show that sharks do not have an intention to harm when approaching a person from behind; they do not have a mean streak, and their goal is not to sneak up and bite but take advantage of approaching the blind side of a person to remain unnoticed.

1.4. What do bite rates tell us?

The study of shark-human interactions does not only deal with the actual communication between sharks and humans in different scenarios but also endeavors to understand where and when bites occur [33, 50, 51].

Between 1994 and 2009, 546 shark bite incidents occurred along the coast of the continental United States (some incidents were excluded where direct provocation by the person triggered the bite, e.g., grabbing a shark). Around 2/3 of these incidents occurred in Florida (365 cases), followed by the Carolinas (83 cases) and California (50 cases).

Although there is a plethora of information on shark bites and the places where they occur, they often only reflect single events [52] or focus on bite numbers. The former often uses a more detailed forensic approach, which can be of great value (see later paragraph); the latter is often rather flawed since the absolute bite number is usually not put into its proper context. The obvious shortcoming of such numbers is that they cannot be easily compared, except for the actual difference since no connection is made to the numbers of people on the beach at the time. To overcome this shortcoming, we defined the term "bite rates," which reflects the ratio between the total annual bites for a certain region to the yearly estimated beach attendance for that region [33]. For our studies, the beach attendance data were taken from the estimates determined by the United States Lifesaving Association (USLA), a not-for-profit association of professional lifeguards. It is understood that these numbers represent the beach attendance, and not how many people were actually in the water at a given time, but the data were sufficient as a population proxy of people entering the water. Although beach attendance numbers are fairly well established, there are some areas for example in Florida where shorelines are not observed by lifeguards, and thus no numbers were available. In order to overcome such shortcomings, we approximated beach populations by using data from adjacent beaches where populations were known. We then applied a spatial scan statistical tool, called

SaTScan [53–55], classified each incident and assigned it to a specific area (a unit) within a county. Then, we analyzed the data in a three-dimensional and two-dimensional manner, with and without the aspect of time. The approach identified primary and secondary regional clusters, which showed significantly higher or lower bite rates than average.

The obvious advantage of bite rate measurements over total bite numbers is that we are able to accurately look for common denominators. This is especially helpful where high and low clusters are adjacent to each other.

The coast of Florida is well known as the region with the most bites annually, and this is especially pronounced on the shoreline of Volusia County, which is also labeled as the "shark attack capital of the world." However, despite the superlative label, the probability of getting bitten is still more than three times lower than for Charleston County in South Carolina [50].

Although different environmental factors come into play when shark bites are considered, we also have looked at more general phenomena, such as the cycle of the moon [56]. It has been suggested that lunar cycles play a role in influencing the number of bites; during the periods of the full and new moon, the number of bites would be increased. We compiled all the bite data from 2002 to 2011 and compared their occurrence with the respective lunar cycle. In order to illustrate any possible correlations more clearly, we used the number of bites on the very day of the full and new moon within the range of ±3 days of that date. The results clearly showed that there is no correlation between the number of reported shark bites and the stage of the lunar cycle. Does this mean that the lunar cycle and its influence on tides do not have an effect on the sharks? Not at all, but the effects are likely subtler, and our data only suggest that it is just not related to shark bite incidents. Although there is no obvious correlation between shark bites and lunar cycles, the question remains why people had the idea in the first place. There is more light during a full moon, thus it seems plausible that this could facilitate the interaction between a surfer or swimmer and a shark during nighttime hours, but the opposite would be true during the new moon phase, which then would need another explanation.

There are numerous factors that have been previously mentioned, which could increase a shark's curiosity, like the water's low visibility encouraging a shark to get closer, a person's struggling motion or fishing close to where people swim, to just name a few. Similar to the lunar cycle idea discussed above, it is prudent to look at each suggestion individually and make sure that the source of the information is legitimate.

Probably, the most obvious and simplest anthropogenic factor related to bites is the constantly increasing number of human settlements close to shorelines and thus greater numbers of people entering the water. Although increasing numbers of humans will tend to increase the number of incidents, the noise created could also have the opposite effect and repel the sharks. A reduction in incidents could also be triggered by other factors such as eutrophication [57, 58] or harmful algae blooms infecting and limiting the shark's food base. All these potential influences need to be thoroughly investigated in well-planned studies.

Factors attracting sharks to investigate a human may also imply a motivational change in the animal. Therefore, realizing what keeps a shark in a certain area or what lured it in is an important area of study. In most cases, an attractant lowers the hesitation threshold in a shark, motivating it to come closer. This per se is not dangerous, but the situation could get out of hand should the person not pay enough attention or react in the wrong way.

1.5. Comparison of different shark species

Environmental influences facilitate the outcome of an interaction between sharks and humans, but so do biological ones. The latter could be related to, e.g., mating, where a shark sees a person as a potential competitor, migration routes [51], where sharks following fish schools which engulfs a swimmer or surfer, and so on. But there could also be the possibility that smaller sharks show up or remain in some shallower areas to avoid being hunted by larger sharks. Despite the rather obvious connection between a small shark and a shallow area, the general causation of a larger shark's presence cannot always be definitively pinpointed. Even more so, the species identification may not be possible, as is often the case along Florida's shores [42]. As we already mentioned above, there are probably four species of the genus *Carcharhinus* that are responsible for most of the incidents along Florida's east coast. Despite the fact that each species can clearly be identified in a proper setting, its general physique and coloration make it hard for everyone, should the animal not clearly be seen or then only for a fraction of a second. That is not only true for their general appearance but also for the wound patterns they cause. The only species that has a clearly distinct pattern of tooth imprints is the bull shark. But even for that species, a superficial bite from a smaller specimen is hard to distinguish from others. This uncertainty makes analysis of why the responsible sharks mill around in certain areas at particular times challenging because species-specific attributes cannot be used. A much clearer picture with regard to species identification can be gained along the Pacific coast, where most incidents are caused by white sharks, *Carcharodon carcharias* [51]. This is a well-studied species in these waters [59], and many of its biological traits, including hunting tactics, migration patterns and so on, are well known. These traits could explain the high and low incident clusters and so permit being proactive and closing beaches during times when the number of sharks is likely increased [51].

The more we know about sharks, their migration routes, their hunting and nursery grounds, the easier it is to predict where possible conflicts may happen.

1.6. On the most exposed human water activity

Our results clearly showed that of all water-based activities, surfing has the highest exposure to contact with sharks [33]. Several factors seem to underlie this observation. Probably, the most prominent factor is the duration surfers stay in the water compared to other beachgoers. However, their frequent close proximity to sandbars may also have an effect although this needs more research. The reason for surfing close to sandbars is obvious; however, the attraction for sharks to sandbars is less clear. Of course, there is the oxygen-enriched water, caused by the breaking of the waves, which would allow sharks to increase their oxygen intake while having to swim less, but it could also be because more plankton accumulates within these breakers, thus attracting more fish. Due to the low visibility in these zones, the chances are that a surfer who falls off his or her board or is paddling may startle a close by shark, resulting in a possible bite. Examination of the wounds of surfers from the East coast of Florida for example shows that the wounds sustained are mostly superficial in nature, indicating exploration as the main underlying motivation, which is the same result as for our study of white sharks biting surfers in California and Oregon [7].

Conflicts between surfers and sharks certainly need to be examined with more scrutiny. One of the most intriguing questions is whether sharks, white sharks and *Carcharhinus* species alike, interpret a surfer to be an animate or inanimate object [7].

1.7. Reading bite wounds

Whenever a bite incident occurs, the best indication of a shark's motivation lies in the detailed analysis of the wound, since possible eyewitness accounts or even descriptions by the victims are often biased. Trying to understand a wound and its creation follows the general methods of forensic analysis [8, 9]. One of the most least appreciated facts is that a wound is not just caused by the shark itself but the motion of the person as well. Most incidents are caused by the shark's curiosity to get further information of what a human being could be. A rather gentle grabbing by the jaws is sufficient to give a shark plenty of information. But it is this low-pressure bite that allows a person to jerk away his or her arm, or leg, thus causing a secondary wound.

Exploration is by far the most often frequently observed motivation for shark bites, followed by stress/startling, competition and provocation. Further causes are object play, target practice or a combination of different motivations labeled as "pattern compensation" [7]. As it is indicated that a variety of mental states underlie these bites, ranging from curiosity to self-defense. However, there are two possible motivations for shark bites that have never been observed: hunger and striking back, the deliberate intention to harm a person.

As described, exploration is the most prominent motivation underlying shark bites. Although these wounds are mostly very superficial, due to the low jaw pressure applied, the secondary wounds, caused by the reaction of the victim, can be rather severe and even fatal, should for example an artery be nicked. One of the main flaws when talking about bite severity is to use the category of "fatal." This outcome is the only one that does not properly describe the physical damage a bite caused. Although a massive bite could happen that leads to the death of the victim, a less severe injury that severs an artery can have the same outcome. Where the first scenario uses a lot of bite strength, the second one can be caused by a rather superficial bite to an area where an artery lies close to the skin's surface. Care should be taken in bite analysis and reconstruction, because wounds are not always what they appear to be, and this is especially true where surfers and sharks are involved. A bite can be caused through exploration which appears to be the case in at least two-thirds of all incidents [7]. Such a motivation might not be apparent in the case of a more intense wound, which could be caused, for example, when the lower teeth get stuck in the underside of the board, and the shark causes further damage as it tries to free itself.

Although we are starting to get a clearer picture why shark bites happen, some cases lack any obvious causative factors and seem to be a fluke of nature, despite that an incident is always caused by several factors what is also called "constellation of factors" [7].

Although the factors that cause bites are not yet fully known, we are starting to get a more complete picture. This raises the question why prevention of incidents, based on this emerging picture, is still in its infancy despite this knowledge, and the only solutions for beach protection seem to remain shark nets and drum lines, which are harmful to the sharks and other marine life [60, 61]. One problem is that there is a severe lack of awareness of which areas should be avoided for any water sport activity. To a nontrained eye, one beach may look like any other beach but that is far from true. Each beach is a unique environment, and some will be more attractive to sharks than others. But as just mentioned above, it is the "constellation of factors" that is crucial. A beach in itself may not cause a problem but should there be a fishing pier close by, it becomes a very different story. Fishermen indirectly attract sharks, and so anybody who swims down current of the pier can be put in danger. Whenever a fisherman hooks a fish, it likely releases stress hormones to warn others, as well as feces and body fluids. These components are then carried away with the current; the more concentrated the initial scent, the farther it can be picked up by sharks, as long as there is sufficient current. To the shark, using a search image, the scent represents a struggling fish, thus it looks for the origin of the scent and while doing so, it may cross paths with a beachgoer. Although the shark understands that the person is not a fish, it may still be interpreted as the source of the scent. Should the person also splash around and the emitted frequencies be similar to a struggling fish, the shark's interest could be more than just peaked. This potential scenario could play out quite often along Florida's beaches, where fishing piers are very common. The fact that more incidents do not occur, despite this constellation of factors is likely due to the sharks rather high threshold to approach an unfamiliar object. To avoid or at least reduce the likelihood of such a scenario, one of at least two regulations should be implemented. First, swimming should only be allowed up current of an active fishing pier, or second, the hours of fishing from a pier should be limited to early morning and late afternoon hours. Whatever restriction is chosen does not really matter as long as the two activities, fishing and swimming, are strictly separated.

1.8. Twist, wiggle, and the lowering of the pectoral fins

An understanding of the body language of sharks would enable improved outcomes in shark-human interactions. For example, probably the best-known aspect of a shark's expression of discomfort, when in the vicinity of a human being, is the lowering of its pectoral fins and hunching, the arching of its back [2]. However, this rather distinctive form of posturing has never been properly investigated. The lowered pectoral fins are most likely wrongfully misinterpreted as an expression of a threat. The likely cause seems to be the opposite; a shark feels threatened by the presence of the person. It is uniformly accepted that, whenever a shark lowers its pectorals, it does so in close vicinity of the person, mostly close to its "inner circle" [12] but never farther away. Should a shark feel the need to threaten a person, for whatever reason, it could also do so when farther away. However, that never seems to be the case. The phenomenon of lowering pectoral fins can be looked at in two ways: first, the lowering motion itself, and second, the final position of lowered pectoral fins. We observed on many occasions that the lowering motion comes into play when the shark needs to pivot or turn close to a person in a tight space. In this case, it lowers its both pectoral fins very quickly and twists one of them in such a way that it can create a left or right turn but hardly ever a downward motion. Whenever the pectorals are used in such a way, it is always done very rapidly and is called a pectoral burst [2]. When the movement is done more slowly, it seems to help maneuvering by increasing the lateral surface of a turning shark. In both cases, the pectoral fins are used for steering. Although one could argue that the movement still appears to express the displeasure of the moving shark, but the opposite still seems more plausible because it always moves the animal away from a person. A display that offers a very distinct advantage from a maneuvering viewpoint does not likely serve the purpose of a threatening display. Such a display would be conspicuous and goal oriented, directed toward the targeted object and not away from it.

Hunching on the other hand could have the assumed purpose of a threat display when directed against another shark, but this also has to be questioned when humans are targeted. Whenever divers seem to see this display, it is most likely caused by sharksuckers, *Echeneis naucrates*, attached to a shark's body irritating its sensory organs. One of these sharksucker-induced behaviors called wiggling [23] looks very similar to the initial posturing, identified by Johnson and Nelson [62] that was then declared as a threat display. Since sharksuckers are very commonly associated with sharks [23–25], it can be assumed that persons being witnessing these patterns may very well misinterpret what they see. However, the actual threat display described by Johnson and Nelson [62] with grey reef sharks, *Carcharhinus amblyrhynchos*, on this one occasion, has never been scientifically mentioned again in another setting or with another species. This seems rather odd, considering the number of people that encounter sharks in a variety of circumstances and surroundings, such as the tight spaces within reefs that seem to have facilitated the original description [62]. Could it be that this description was indeed a misinterpretation, and that the posturing observed was merely a fluke of nature? We cannot say for sure, but considering the frequency of shark-human encounters, we should be able to observe this type of posturing more often. It is up to the reader to make a decision what this display truly means or what triggers it, but looking at the original work and the two accompanying pictures reveal that in both cases, sharksuckers were attached to the pelvic areas of the shark [62].

We have shown in our previous studies that sharksuckers trigger a variety of effects [23–25], which have a wide spectrum, from very subtle to rather grotesque. Should a shark try to chafe the sharksucker off on the sea bottom [63], the meaning is rather obvious, but there is not always suitable bottom to do so, or the surface may not be close enough to jump out of the water and fall back with the targeted area first [64]. In such cases, the shark tries to get rid of a sharksucker by pushing or pulling the skin at the location of the attached teleost to loosen its suction, which can look rather intimidating considering the shark has to keep swimming to do this.

1.9. Yawning and other jaw gapes

The more complete the ethogram of sharks, the better the understanding of its presence close to a person. There is an obvious fascination in a shark's gape and its teeth, and this is at the forefront of many descriptions given when divers encounter them. One of the most memorable displays is gaping of the jaw, commonly called yawning [24]. Although a true maintenance behavior, to an uninformed person it can look rather intimidating should a shark swim in their general direction, while performing this type of gape. Whenever a shark tries to gouge a piece out of a larger prey or carcass, its upper jaw is often everted. While doing so, a variety of muscles, tendons and ligaments, between this jaw and the connected brain capsule, are put to use. Due to the number and complexity of tendons and ligaments, the repositioning of the upper jaw into its resting position below the brain case can then be rather difficult, and not all the ligaments may be repositioned correctly again. Such misplacement seems to trigger this form of gaping, which often takes 10 times longer than a real bite [24]. The slowed down procedure is likely used to increase the chance of a correct placement. On rare occasions, the first gape, or yawn, is ineffective and the shark tries again by using a different speed.

In addition to this maintenance behavior, there is another type of gape that is occasionally seen among lemon sharks, *Negaprion brevirostris*, when lying on the sea bottom. In these cases, a shark opens its mouth as a signal to get cleaned between its upper teeth by sharksuckers [25]. This symbiosis has an ancient evolutionary history and was probably initiated back in the Eocene-Oligocene period, when sharksuckers first appeared [65, 66]. The interesting part of this cleaner-client relationship is that both the sharksucker and the shark can initiate a cleaning bout. The shark either opens its mouth and awaits a sharksucker, or the sharksucker swims up and down in front of the shark's eye in a dance-like manner to trigger a response, or can use a tactile response for the same reason. For a shark to be aware that food is stuck between its teeth, a sensory mechanism needs to be in place. However, it is unknown how this may work.

Likewise, it is not yet understood why sharks need to have their upper teeth cleaned at all, considering that they continually replace their teeth every 1–4 or 5 weeks [67–69]. Assuming that food resting between the teeth starts to rot, and that a broken tooth exposes blood vessels, the possibility of infection could then justify cleaning.

1.10. Shark-human interactions: a tool for shark conservation

The previous sections reflect either new explanations for old assumptions (e.g., that sharks do not have a threatening posture in front of humans) or explain commonly seen behavioral patterns of sharks (e.g., that pectoral fin lowering is used for a change of direction or increased maneuverability). In any case, the results facilitate the better understanding of the behavior of sharks in the vicinity of humans.

Shark behavioral patterns can be triggered by a multiple of triggers, such as human presence, nonhuman irritations, low visibility and others. Whenever explaining shark behavior, it is crucial to always describe the circumstances in which the behavior occurs and to highlight its role within the context of shark-human interactions. Each piece of knowledge helps to complete the picture we have of sharks. The better we understand these animals, how they function and what triggers an action or a reaction, the more fruitful our interaction with them as a species will ultimately be.

Although people feel afraid of a shark in front of them, they are actually just projecting their numerous fears into the animal; thus, a better understanding of the underlying meaning and influences of certain shark behaviors will help to keep a person calm. Human fear among sharks comes from a lack of understanding of the situation, how it develops and most importantly what to do should a reaction be necessary. Therefore, it is of critical importance to "get to know" sharks and that includes being among them. Although science teaches us various techniques and procedures when dealing with different kinds of animals, one should always take time to observe them and let situations take place. "Getting to know" a shark or "getting a feeling" for them is crucial.

We are at a point in our research where we can offer answers to many shark-related behavioral questions, helping to create a far different animal to the one portrayed in JAWS. As already mentioned, it is an old adage that people do not protect what they are afraid of, thus by showing how sharks truly are and understanding how they behave, we may be able to raise the kind of awareness that leads to long-lasting protection of sharks and reduced incidents in which humans are harmed. Such protection is paramount if we want to save our oceans.

1.11. Future research in shark-human interactions

The examples in this chapter show how experimental approaches are uncovering the different aspects of shark-human interactions. This chapter also shows how all this information together can support the same goal: to better understand sharks. Some shark species have been brought to the brink of extinction largely due to human ignorance and the fact that people did not care enough about them due to the bad reputation these creatures had. Attitudes need to be changed on a worldwide level and that can only be accomplished by making sharks better understood. Although we are currently tackling a variety of questions within the experimental shark-human interaction field, some areas are still largely untouched.

Sharks are the most unknown and still the most abundant large predators in our oceans. Their presence in the marine realm demands that we get to know them better before it is too late. Any research helps, but the studies most able to change people's attitude and perception about sharks are those focusing on aspects of shark-human interactions.

Author details

Erich Ritter* and Raid Amin

*Address all correspondence to: eritter@uwf.edu

University of West Florida, Pensacola, FL, USA

References

[1] Ritter E, Lutz K, Levine M. When humans and sharks meet. In: Olsson F, editor. New Developments in the Psychology of Motivation. New York: Nova Biomedical Books; 2008. pp. 45-52

[2] Ritter, E. Shark-Human Interaction. Situations Findings Recommendations. Pensacola, FL: SharkSchool Publishing; 2012. p. 165

[3] Tougias T. When Man is the Prey. New York: St. Martin's Griffin; 2007. p. 397

[4] May N, Willis C. Shark: Stories of Life and Death from the World's Most Dangerous Waters. Cambridge, MA: Da Capo Press; 2002. p. 368

[5] Allen TB. Shark Attacks: Their Causes and Avoidance. Gilford, CT: The Lyons Press; 2001. p. 312

[6] Ritter E. Sharks-Mistaken identity? In: Bekoff M, editor. Encyclopedia of Animal Behavior. Westport: Greenwood Press; 2004. pp. 963-964

[7] Ritter EK, Quester A. Do white shark bites on surfers reflect their attack strategies on pinnipeds? Journal of Marine Biology. 2016. DOI: 10.1155/2016/9539010 2016

[8] Ritter EK, Levine M. Bite motivation of sharks reflected by the wound structure on humans. American Journal of Forensic Medicine and Pathology. 2005;**26**:136-140

[9] Ritter EK, Levine M. Use of forensic analysis to better understand shark attack behaviour. Journal of Forensic Odonto-Stomatology. 2004;**22**:40-46

[10] Ritter E, Amin R. Are Caribbean reef sharks, *Carcharhinus perezi*, able to perceive human body orientation? Animal Cognition. 2014;**17**:745-753

[11] Ritter E, Amin R. Effect of human body position on the swimming behavior of bull sharks, *Carcharhinus leucas*. Society and Animals. 2012;**20**:225-235

[12] Ritter E, Amin R. A study of stealth behavior in the proximity of divers. Open Journal of Animal Sciences. 2015;**5**:224-228

[13] Philpott R. Why sharks may have nothing to fear than fear itself: An analysis of the effect of human attitudes on the conservation of the white shark. Colorado Journal of International Environmental Law and Policy. 2002;**3**:445-472

[14] Boissonneault MF, Gladstone W, Scott P, Cushing N. Grey nurse shark human interactions and portrayals: A study of newspaper portrayals of the grey nurse shark from 1969-2003. Electronic Green Journal. 2005;**1**:1-21

[15] Muter BA, Gore ML, Gledhill KS, Lamont C, Huveneers C. Australian and U.S. news media portrayal of sharks and their conservation. Conservation Biology. 2012;**27**:187-196

[16] Baldridge HD. Shark attack: A Program of Data Reduction and Analysis. Contributions from the Mote Marine Laboratory. Tampa, FL; 1974. p. 98

[17] Driscoll JW. Attitude toward animals: Species ratings. Society and Animals. 1995;**3**:139-150

[18] Woods B. Beauty and the beast: Preferences for animals in Australia. Journal of Tourism Studies. 2000;**11**:25-35

[19] Thompson TL, Mintzes JJ. Cognitive structure and the affective domain on knowing and feeling in biology. International Journal of Science Education. 2002;**24**:645-660

[20] Houston MJ, Bruskotter JT, Fan D. Attitudes toward wolves in the United States and Canada: A content analysis of the print news media, 1999-2008. Human Dimensions of Wildlife. 2010;**15**:389-403

[21] Stevens JD, Bonfil R, Dulvy NK. The effects of fishing on sharks, rays, and chimaeras (chondrichthyans), and the implications for marine ecosystems. ICES Journal of Marine Science. 2000;**57**:476-494

[22] Myers RA, Baum JK, Shepherd TD, Power SP, Peterson CH. Cascading effects of the loss of apex predatory sharks from a coastal ocean. Science. 2007;**5820**:1846-1850

[23] Ritter E. Analysis of sharksucker, *Echeneis naucrates*, induced behavior patterns in the blacktip shark, *Carcharhinus limbatus*. Environmental Biology of Fishes. 2002;**64**:111-115

[24] Ritter E. Mouth gaping behavior in Caribbean reef sharks, *Carcharhinus perezi*. Marine and Freshwater Behaviour and Physiology. 2008;**41**:161-167

[25] Ritter EK, Amin R. Mouth cleaning of lemon sharks. Negaprion brevirostris, by sharksuckers, *Echeneis naucrates*. Copeia. 2016;**104**:728-733

[26] Seligman M. Phobias and preparedness. Behavior Therapy. 1971;**2**:307-320

[27] Seligman M. Competing theories of panic. In: Rachman S, Maser JD, editors. Panic: Psychological Perspectives. Hillsdale, NJ: Lawrence Erlbaum Associates; 1988. pp. 321-320

[28] Krop H, Krause S. The elimination of a shark phobia by self administering systematic desensitization: A case study. Journal of Behavior Therapy and Experimental Psychiatry. 1976;**7**:293-294

[29] Ulrich RS. Biophilia, biophobia, and natural landscapes. In: Kellert SR, Wilson EO, editors. The Biophilia Hypothesis. Washington, D.C.: Island Press; 1993. pp. 73-137

[30] Kellert SR. The Value of Life. Washington, D.C.: Island Press; 1996. p. 282

[31] Tricas TC, McCosker JE. Predatory behavior of the white shark (*Carcharodon carcharias*), with notes on its biology. Proceedings of the California Academy of Science. 1984;**43**:221-238

[32] Lea R, McCosker J. Shark-human interaction in the eastern North Pacific: An update and analysis [Internet]. 1993. Available from: http://archive.rubicon-foundation.org/xmlui/handle/123456789/6329

[33] Amin R, Ritter E, Kennedy P. A geospatial analysis of shark attack rates for the east coast of Florida: 1994-2009. Marine and Freshwater Behaviour and Physiology. 2012;**45**:185-198

[34] Casey JG, Pratt HL. Distribution of the white shark, Carcharodon carcharias, in the western North Atlantic. Memoirs of Southern California Academy of Science. 1985;**9**:2-14

[35] Klimley AP. The areal distribution and autoecology of the white shark, Carcharodon carcharias, off the west coast of North America. Memoirs of Southern California Academy of Science. 1995;**9**:15-40

[36] Hussey NH, McCann G, Cliff G, Dudley FJ, Wintner SP, Fisk AT. Size-based analysis of diet and trophic position of the white shark, Carcharodon carcharias, in South African

waters. In: Domeier M, editor. Global Perspective on the Biology and the Life History of the White Shark. Boca Raton: CRC Press; 2012. pp. 27-49

[37] Ainley DG, Henderson RP, Huber HR, Boekelheide RJ, Allen SG, McElroy T. Dynamics of white shark/pinniped interactions in the Gulf of the Farallones. Memoirs of Southern California Academy of Science. 1995;**9**:109-122

[38] McCosker JE. White shark attack behavior: Observations of and speculations about predator and prey strategies. Memoirs of the Southern California Academy of Science. 1995;**9**:123-135

[39] Martin RA, Hammerschlag N, Collier RS, Fallows C. Predatory behaviour of white sharks (*Carcharodon carcharias*) at Seal Island, South Africa. Journal of the Marine Biological Association of the United Kingdom. 2005;**85**:1121-1135

[40] Shimada K. The relationship between tooth size and total body length in the white shark, *Carcharodon carcharias* (Lamniformes: Lamnidae). Journal of Fossil Research. 2002;**35**:28-33

[41] Lowry D, de Castro ALF, Mara K, Whitenack LB, Delius B, Burgess GH, Motta P. Determining shark size from forensic analysis of bite damage. Marine Biology. 2009;**156**:2483-2492

[42] Shark Research Institute. Global shark attack file incident log [Internet]. Available from: http://www.sharkattackfile.net/incidentlog.htm [Accessed January 1, 2017]

[43] Baldridge HD. Shark aggression against man: Beginnings of an understanding. California Fish and Game. 1988;**74**:208-217

[44] Collier RS. Recurring attacks by white sharks on test-subjects at two Pacific sites off Mexico and California. Environmental Biology of Fishes. 1992;**33**:319-325

[45] Byard RW, Gilbert JD, Brown K. Pathologic features of fatal shark attacks. American Journal of Forensic Medicine and Pathology. 2000;**21**:225-229

[46] Call J, Bräuer J, Kaminski J, Tomasello M. Domestic dogs (*Canis familiaris*) are sensitive to the attentional state of humans. Journal of Comparative Psychology. 2003;**117**:257-263

[47] Gácsi M, Miklósi A, Varga O, Topál J, Csáni V. Are readers of our face readers of our minds? Dogs (*Canis familiaris*) show situation-dependent recognition of human's attention. Animal Cognition. 2004;**7**:144-153

[48] Kaminski J, Call J, Tomasello M. Body orientation and face orientation: Two factors controlling apes' begging behavior from humans. Animal Cognition. 2004;**7**:216-233

[49] Pack AA, Herman LM. Bottlenosed dolphins (*Tursiops truncatus*) comprehend the referent of both static and dynamic human gazing and pointing in an object-choice task. Journal of Comparative Psychology. 2004;**118**:160-171

[50] Amin R, Ritter E, Wetzel A. An estimation of shark attack risk for the North and South Carolina coast line. Journal of Coastal Research. 2014;**31**:1253-1259

[51] Amin R, Ritter E, Cossette L. An investigation of shark density and attack rates in California. Journal of Environment and Ecology. 2013. DOI: 10.5296/jee.v3i1.2700

[52] Clua E, Reid D. Features and motivation of a fatal attack by a juvenile white shark, *Carcharodon carcharias*, on a young male surfer in New Caledonia (South Pacific). Journal of Forensic and Legal Medicine. 2013;**20**:551-554

[53] Kulldorff M. SaTScan v. 8.0: Software for the Spatial and Space–Time Scan Statistics [Internet]. 2009. Available from: http://www.SaTScan.org

[54] Kulldorff M, Heffernan R, Hartman J, Assuncao RM, Mostashari FA. Space–time permutation scan statistic for the early detection of disease outbreaks. PLoS Medicine. 2006;**2**:216-224

[55] Kulldorff MA. Spatial scan statistic. Communications in Statistics. Theory and Methods.1997; **26**:1481-1496

[56] Ritter E, Amin R, Zambesi A. Do lunar cycles influence shark attacks? Open Fish Science Journal. 2013;**6**:71-74

[57] Anderson DM, Glibert PM, Burkholder JM. Harmful algal blooms and eutrophication: Nutrient sources, composition, and consequences. Estuaries. 2002;**25**:704-726

[58] Andersson T, Förlin L, Härdig J, Larsson A. Physiological disturbances in fish living in coastal water polluted with bleached kraft pulp mill effluents. Canadian Journal of Fisheries and Aquatic Sciences. 1988;**45**:1525-1536

[59] McCosker JE, Lea RN. White shark attacks upon humans in California and Oregon, 1993-2003. Proceedings of the California Academy of Sciences. 2006;**57**:479-501

[60] Atkins S, Cliff G, Pillay N. Humpback dolphin bycatch in the shark nets in KwaZulu-Natal, South Africa. Biological Conservation. 2013;**159**:442-449

[61] Cliff G, Dudley SFJ. Reducing the environmental impact of shark-control programs: A case study from KwaZulu-Natal, South Africa. Marine and Freshwater Research. 2011; **62**:700-709

[62] Johnson RH, Nelson DR. Agonistic display in the gray reef shark, *Carcharhinus menisorrah*, and its relationship to attacks on man. Copeia. 1973:76-84

[63] Ritter E. Use of sand ripples to enhance chafing in Caribbean reef sharks (*Carcharhinus perezi*) and blacktip sharks (*Carcharhinus limbatus*). Bulletin of Marine Science. 2011; **87**:413-419

[64] Ritter EK, Brunnschweiler JM. Do sharksuckers, Echeneis naucrates, induce jump behaviour in blacktip sharks, *Carcharhinus limbatus*? Marine and Freshwater Behaviour and Physiology. 2003;**36**:111-113

[65] Micklich N. New information of the fishfauna of the Frauenweiler fossil site. Italian Journal of Zoology. 1998;**65**:169-184

[66] O'Toole B. Phylogeny of the species of the superfamily Echeneoidea (Perciformes: Carangoidei: Echeneidae, Rachycentridae, and Coryphaenidae), with an interpretation of echeneid hitchhiking behaviour. Canadian Journal of Zoology. 2002;**80**:596-623

[67] Moss SA. Tooth replacement in the lemon shark, Negaprion brevirostris. In: Gilbert PW, Mathewson RF, Rall DP, editors. Sharks, Skates, and Rays. Baltimore: John Hopkins Press; 1967. pp. 319-329

[68] Reif WE, McGill D, Motta P. Tooth replacement rates of the sharks *Triakis semifasciata* and *Ginglymostoma cirratum*. Zoologisches Jahrbuch der Anatomie. 1978;**99**:151-156

[69] Luer CA, Blum PC, Gilbert PW. Rate of tooth replacement in the nurse shark, *Ginglymostoma cirratum*. Copeia. 1990:182-191

Section 2

Biodiversity, Phylogenetic Systematics in Chondrichthyes

Chapter 5

Deep-Water Sharks, Rays, and Chimaeras of Brazil

Getulio Rincon, Rodrigo Cordeiro Mazzoleni,
Ana Rita Onodera Palmeira and Rosangela Lessa

Additional information is available at the end of the chapter

http://dx.doi.org/10.5772/intechopen.69471

Abstract

The deep-water fishery in Brazil is currently in expansion due to depletion of most neritic economic species. This increasing deep-water effort brings concern on the bycatch impact, its specific composition, the need for capture's evaluation and development of bycatch reduction devices. The impact is particularly aggressive on deep-water elasmobranchs, which have an extreme ecological k-strategy due to their reproductive constraints (lower fecundity and late first maturity age). Scientific deep-water surveys and intensive research programs (REVIZEE) along the past decade indicate that Brazilian elasmobranch diversity is higher than previously imagined. However, the deep-water fishery threatens this poorly known community of sharks and rays on the Brazilian continental slope as they become bycatch of a fast-growing and uncontrolled fishery. The recent study case of the monkfish (*Lophius gastrophysus*) fishery dynamics, well presented and discussed by the Brazilian scientific community, provided evidence of the need of bycatch-specific monitoring programs and fast-response fishery regulations. The present work discusses the Brazilian deep-water elasmobranch bycatch problem under the light of its biological diversity and completely unknown population status. Suggestions and management considerations are presented in order to coordinate and manage the establishment and growth of this deep-water fishery in Brazil.

Keywords: elasmobranch, fish, continental slope, biodiversity, demersal, Chondrichthyes

1. Introduction

The development and use of deep-water fishing gears (bottom trawl, bottom longline, gillnets and pot traps) along the Brazilian continental slope have increased significantly after the depletion of the majority of valuable coastal fishing resources over the past decades [1–5]. However, this search for new fishing grounds and change of fishing habits sometimes occurs suddenly in a "gold rush" and spasmodically as old resources collapse and new resources are discovered and/or extreme catch variations are reported [3, 6–10]. This pioneer new fishery is characterized by excessive investments encouraged by regulations and subsidies [3, 11–13], leading great part of the local fleet to target mostly a single valuable species for a distinct foreign market. As a result, this particular resource sustains high catches for a short time, usually 2 or 3 years, followed by a complete economic collapse [6, 9]. This fishing behavior and voracity led researchers to find themselves in a cat-and-mouse game where fishing resources are exploited before the scientific community, which is aware of it or even far from an available population management data base [7]. Probably even more serious is the capture of a large number of bycatch species, especially those of smaller size, highly represented and particularly with a population dynamics sensitive to fishery mortality, such as sharks, rays, and sea birds [2, 14–16].

The Brazilian deep-water elasmobranch diversity is poorly known only by occasional records based on commercial and scientific surveys, mostly as part of the recent Brazilian Program for Evaluation of the Living Resources in the Exclusive Economic Zone (REVIZEE). A recent increase of deep-water research along the Brazilian continental slope in the past decades [1995–2007] has revealed a great number of new species or records of elasmobranch fishes [17–31], which is changing the marine biogeographic relevance of the Neotropical realm for this group of vertebrates. Nineteen species were recorded for the first time in Brazil or described for the continental slope habitat over the past 15 years. The list of species known to occur on the Brazilian continental slope increases significantly every year and suggests that the total number of species is still beyond the current knowledge [17, 18, 32–36]. Several recent records were at the genus level (*Somniosus, Dalatias, Cirrhigaleus, Centrophorus, Apristurus, Galeus, Parmaturus, Pseudotriakis, Mitsukurina, Benthobatis, Dactylobatus, Cruriraja, Breviraja,* and *Malacoraja*) and four new families were recorded for the first time (Anacanthobatidae, Pseudotriakidae, Centrophoridae, and Mitsukurinidae), [*q.v.* 17, 27, 37–39] (**Figure 1**). Currently, several deep-water species are under description or under review process (e.g., *Scyliorhinus* sp. B and *Dipturus* sp. A and sp. B, *Parmaturus* cf. *campechiensis* Springer 1979). These findings are evidence of unevaluated deep-water species diversity and bring concern on the possible near-future development of deep-water fisheries where scientific research is still badly needed.

Recent deep-water scientific trawls off the north and southern coasts of Brazil have revealed several unexploited stocks of prawns such as *Aristeus antillensis* Milne-Edwards and Bouvier 1909 and *Aristaeopsis edwardsiana* (Johnson, 1868), and crabs of the genus *Chaceon* [38, 40–42]. Additionally, as a result, the local fleet has considered adapting the necessary gear for deep-water bottom trawl in search of these valuable crustaceans, but mostly chartered fleet landings have operated and the fishery developed mainly from 2002 to 2009 (*q.v.* [42] for a review). The area is part of the known range of distribution of species such as: *Schroederichthys tenuis* Springer 1966, *Scyliorhinus ugoi* Soares, Gadig and Gomes 2015,*Breviraja spinosa* Bigelow and Schroerder 1950, *Gurgesiella atlantica* (Bigelow and Schroeder 1962), *Dipturus* spp., *Rajella purpuriventralis* (Bigelow and Schroeder, 1962) and the recently recorded goblin shark *Mitsukurina owstoni* Jordan 1898 [17, 38, 39]. According to Asano-Filho et al. [38], about 4% (4278 kg) of the total experimental capture (kg/h) of a series of scientific trawls consisted

http://dx.doi.org/10.5772/intechopen.69471

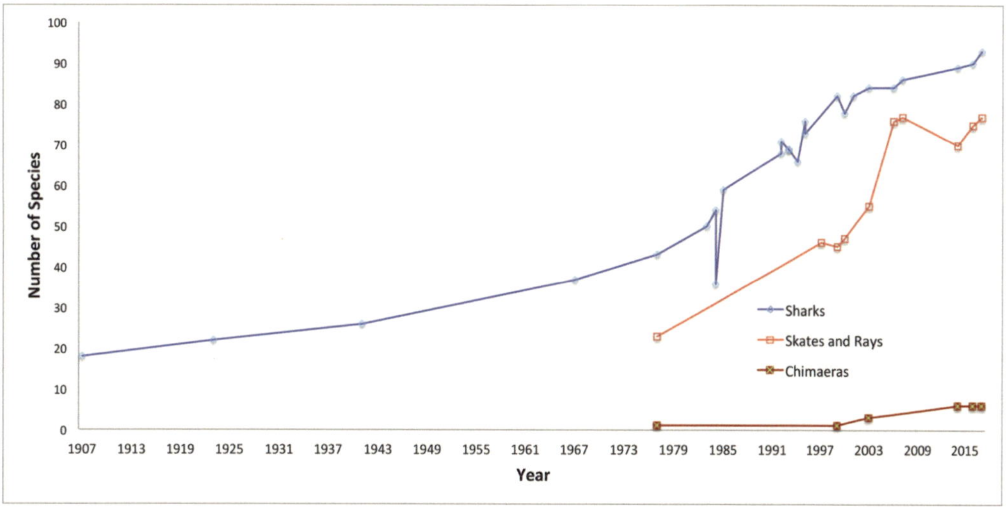

Figure 1. Total number of Chondrichthyes recorded along the time in Brazilian waters. This numeric compilation considered published papers, thesis and species accounts presented in scientific meetings.

of elasmobranchs. However, these authors did not consider the numeric capture per species, and since most of these species listed above are small to medium sized forms (excepting adult *Dipturus* and *Mitskurina*), the impact on the elasmobranch community may be more important than previously supposed by analyzing solely the Capture-Per-Unit-Effort (CPUE) in weight terms.

The monkfish fishery off the southern Brazil was initially established by chartered boats that introduced the deep-water net and trawl fishing technology, quickly followed by the national fleet on a smaller scale [3, 4]. This fishery along with the seasonal deep-water trawl for squids [*Loligo* sp. and *Illex argentinus* (Castellanos, 1960)] emerged fast with surprising landings that reached 4300 metric tonnes of monkfish in 2001 and 2300 tonnes of squid in 2002 [43, 44]. These landings represented a yield increase of up to 1147 and 6597% for monkfish and squid, respectively, in relation to the previous years. The "gold rush" for these resources had just started; although both groups would never reach these levels of catch again. The chartered boat program was reviewed in 2002; mainly because of the influence and research on the monkfish fishery by the scientific crew of the Universidade do Vale do Itajaí-UNIVALI. However, this fishery continues and tends to expand toward the continental slope of the north and northeast coasts of Brazil [45], that brings concern on how fast local researchers will be able to follow fisheries growth and trends.

Here we present a brief review on Brazilian deep-water Chondrichthyes and examine the risk of population collapse of untargeted and poorly recorded deep-water cartilaginous fishes captured by deep-water fisheries, particularly based on the well recorded case of the monkfish [*L. gastrophysus* (Miranda-Ribeiro, 1915)], Argentine squid (*I. argentinus*), and aristeid shrimp (*A. antillensis* and *A. edwardsiana*) fisheries off southern Brazilian coastland available data from the REVIZEE Program.

2. Assembling deep-water information

Deep-water species were considered as those whose captures occurred only, or at least mainly, deeper than 200 m, coincident with the continental shelf edge [46]. This definition also considers the depth distribution in the water column as many species presents a vertical migration in circadian cycles to epipelagic zones at night and deeper waters at day [47–49]. All species identification and nomenclature herein presented followed [50, 51].

All analyzed specimens were captured on the continental slope of Brazil (**Figure 2**) between depths of 130 and 1000 m, by commercial and research vessels between 1991 and 2008 using bottom longline, otter trawl, pot traps, and gillnets. Most of the results from research cruises make part of the Brazilian Program for Evaluation of the Living Resources in the Exclusive Economic Zone (REVIZEE) (1995–1998) or previous local research projects (Fundação Universidade do Rio Grande [FURG] off southern Brazil) (*q.v.* [1, 25, 52–55] for descriptions of capture methodologies). REVIZEE research cruises off the northeast and north coasts used bottom longline and crab pots and otter trawl and longline, respectively ([45] for a longline fishery description). Additional specimens were analyzed from collections at FURG, Universidade Federal Rural de Pernambuco (UFRPE), Universidade Federal do Pará (UFPA), Museu Paraense Emílio Goeldi (MPEG), Museu de Zoologia da Universidade de São Paulo (MZUSP), Universidade Federal da Paraíba (UFPB), Universidade do Vale do Itajaí (UNIVALI), and personal collections of the authors to be institutionally deposited after analysis and publication of results.

Records published in scientific literature, unpublished master and doctoral thesis or simply reported in scientific meetings (abstracts and proceedings) and project reports were also considered as reliable sources of information.

Figure 2. South America and Brazilian coastline as identified by the REVIZEE Program: 1. Northern Coast, 2. Northeastern Coast, 3. Central Coast and 4. Southern Coast.

Bycatch and deep-water fisheries data were based on reports by the UNIVALI research staff (http://gep.acad.univali.br and http://gep.acad.univali.br/?page=estatistica_boletins) and published annual fishery reports by the Brazilian Environmental Agency (IBAMA) (http://www.ibama.gov.br/) up to the year 2011. The gill net monkfish fishery bycatch analysis of [56] provided annual total capture of elasmobranchs and CPUE estimations for 2001.

2.1. Brazilian deep-water elasmobranch bycatch

Elasmobranchs are considered bycatch by most of the deep-water fisheries along the coast and, as a result they are frequently discarded or retained aboard only when the captures of target species are low [15, 43, 56–58]. The only possible retained exceptions are *Galeorhinus galeus* (Linnaeus, 1758), *Squatina* spp., *Squalus* spp., *Mustelus canis* (Mitchell, 1815) and rajids identified as "emplastro" (*e. g., Dipturus* spp., *Atlantoraja castelnaui* (Ribeiro, 1907), *Atlantoraja cyclophora* (Regan, 1903), *Atlantoraja platana* (Günther 1880) and *Rioraja agassizi* (Müller and Henle, 1841)], which may be captured by deep-water fisheries but are considered neritic resources [3, 59]. Eventually, larger species with incipient captures are often retained [*Hexanchus griseus* (Bonnaterre, 1788), *Carcharhinus* spp., *Sphyrna* spp., and others], but they represent less than 5% of all deep-water elasmobranch numeric catches. Perez et al. [58] reported a catch composition of 2.1% of chondrichthyan fishes on the deep-water shrimp fishery in 89 analyzed trawls, and concluded that the whole catch was discarded, which represented at least 1617 specimens of *Etmopterus lucifer* Jordan and Snyder, 1902 and 688 specimens of *Gurgesiella dorsalifera* McEachran and Compagno, 1980. However, some frequently captured small sized species may be unrecorded or misidentified by most of onboard observers (Chartered Fleet Observers Program-Ministry of Agriculture and UNIVALI) and fisheries monitoring programs. These species are mostly treated in generic terms such as "caçonetes" and "cações," or "raias" and "emplastro" (small sharks and rays, respectively) and even hardly considered or identified as bycatch in deep-water fisheries [60]. These groups also include rare and deep-water species such as: *Galeus mincaronei* Soto 2001, *Apristurus* spp., neonates of *Scyliorhinus, Schroederichthys* and *Dipturus* spp., *Etmopterus* spp., *Centrocymnus owstoni* Garman 1906 (previously identified as *C. cryptacanthus* Regan 1906), *Parmaturus* cf. *campechiensis, Cruriraja* sp., *Bathyraja schroederi* (Krefft, 1968) and *Rajella sadowskyii* (Krefft and Stehmann, 1974) (**Table 1**). The lack of precise identification leads to an underestimation of the total capture of the group, especially those highly numbered smaller forms frequently discarded at sea. Some rare and unfamiliar species to most fishery scientists and students were also discarded or unreported along research programs, such as the false catshark *Pseudotriakis microdon* Capello 1868 and the slender catshark *S. tenuis* off northeast and north Brazil, respectively [17, 37, 61].

Scientific surveys usually provide the most reliable source of information to estimate commercial fisheries' bycatch composition as all captured specimens are considered in the analysis and not discarded at sea. However, this bycatch level is frequently much underestimated due to the larger effort and efficiency of commercial vessels in comparison to scientific cruises. The northeastern and central surveys of the REVIZEE program developed experimental bottom longline sets twice a year (November) from 1997 to 1999. These catches presented a chondrichthyan dominance (60% of total catch) [62], and some numbered species suggest that elasmobranchs may become a large component of the bycatch in a future deep-water fishery in the area. Spiny

	Otter Trawl	Gill Net	Longline
Sharks			
Hexanchus griseus		X	X
Heptranchias perlo	X		
Echinorhinus brucus	X		X
Somniosus antarcticus	X		
Somniosus pacificus	X		
Centroscymnus coelolepis	X		
Centroscymnus owstoni	X		X
Zameus squamulosus	X		
Etmopterus pusillus	X		
Etmopterus lucifer	X		
Etmopterus bigelowi	X		X
Etmopterus granulosus	X		
Etmopterus hillianus	X		
Etmopterus gracilispinis	X		
Euprotomicroides zantedeschia	X		
Euprotomicrus bispinatus			
Isistius brasiliensis	X		
Isistius plutodus			
Dalatias licha			X
Squaliolus laticaudus	X		
Squalus acanthias	X	X	X
Squalus bahiensis	X		X
Squalus albicaudus	X		X
Squalus quasimodo	X		X
Squalus lobularis	X		X
Cirrhigaleus asper	X		X
Centrophorus granulosus			X
Deania profundorum	X		
Apristurus parvipinnis	X		
Apristurus profundorum	X		
Apristurus manis	X		
Galeus mincaronei	X		X

	Otter Trawl	Gill Net	Longline
Parmaturus cf. campechiensis	X		
Schroederichthys saurisqualus	X		X
Schroederichthys bivius	X		X
Schroederichthys tenuis	X		
Scyliorhinus haeckelii	X		X
Scyliorhinus ugoi	X		X
Scyliorhinus cabofriensis	X		
Scyliorhinus sp. B	X		
Mustelus canis	X	X	X
Pseudotriakis microdon			X
Mitsukurina owstoni	X		
Squatina sp. A	X	X	X
Squatina sp. B	X		
Rays			
Benthobatis kreffti	X		
Tetronarce puelcha	X	X	
Tetronarce sp. A	X		
Dactylobatus clarkii	X		
Amblyraja frerichsi	X		
Breviraja nigriventralis	X		
Breviraja spinosa	X		
Dactylobatus clarkii	X		
Dipturus bullisi	X		
Dipturus cf. *garricki*	X		
Dipturus leptocaudus	X	X	X
Dipturus mennii	X	X	X
Dipturus teevani	X		
Dipturus sp. A	X	X	X
Dipturus sp. B	X		X
Malacoraja obscura	X		
Rajella fuliginea	X		
Rajella purpuriventralis	X		
Rajella sadowskii	X		
Zearaja chilensis	X		

	Otter Trawl	Gill Net	Longline
Bathyraja brachyurops	X		
Bathyraja multispinis	X		
Bathyraja scaphiops	X		
Bathyraja schroederi	X		
Cruriraja rugosa	X		
Gurgesiella atlantica	X		
Gurgesiella dorsalifera	X		
Chimaeras			
Callorhinchus callorynchus	X	X	
Hydrolagus alberti	X		
Hydrolagus matallanasi	X		
Hydrolagus cf. mirabilis	X		
Harriotta raleighana	X		
Rhinochimaera sp.	X		

Table 1. Deep-water Chondrichthyes of Brazil.

dogfishes and spurdogs [*Squalus* spp. and *Cirrhigaleus asper* (Merrett 1973)] were the main representatives with highest CPUEs (ranging from 1.44 to 5.16 sharks/100 hooks), or a total capture of 795 specimens in six cruises. The dusky smoothhound shark *M. canis* was the second species in abundance with a total capture of 114 individuals and CPUEs that ranged from 0.12 to 2.16 (sharks/100 hooks). The total CPUE for each species is presented in **Table 2** and reveal that only *Squalus* spp., *C. asper*, and *M. canis* could be important economically and all other species would make part of the bycatch if a deep-water fishery develops in the region.

Scientific deep-water otter trawls off Itajaí, state of Santa Catarina, were able to capture about 150 specimens of the blind electric ray *B. kreffti* Rincon, Stehmann and Vooren 2001 in only three trawls [25, 34]. Commercial squid trawlers operating at the same area used to land specimens of the blind electric ray attached or grabbed by the squids and occasional researcher's selections were able to separate up to 300 specimens in one single cruise (Mazzoleni, pers. obs). High CPUE values of rare benthic fishes are strong evidence of an aggregation behavior in specific areas in search of food or for reproductive purposes. The same apparent behavior was also observed in *G. dorsalifera*, a small deep-water skate with evidences of feeding aggregation based on high CPUE values and stomach contents [63–65]. Both species are reported on commercial captures by onboard observers, but frequently not considered as relevant on fisheries management programs. In addition, their aggregation behavior, geographical restriction, and numbered catches by deep-water trawlers make these populations extremely vulnerable [66, 67].

Some species captured by deep-water surveys will be treated individually as follows.

Species/group of species	Total number	Average CPUE (N°/100 hooks)	Depth range
Squalus spp.*	795	2.73	100–500
Mustelus canis	114	0.56	100–400
Scyliorhinus ugoi	16	0.09	100–500
Heptranchias perlo	12	0.11	200–350
Hexanchus griseus	6	0.05	100–300
Echinorhinus brucus	4	0.04	200–250
Centrophorus granulosus	27	1.12	300–350
Pseudotriakis microdon	1	0.01	300–450
Etmopterus spp.**	34	0.18	100–500
Dipturus sp. B	6	0.02	300–500

* *Squalus* spp. = Mostly *S. albicaudus*, but also *C. asper*.
** *Etmopterus* spp. = *E. bigelowi* and possibly *E. pusillus*.

Table 2. Deep-water sharks and rays off the northeast coast of Brazil captured along six cruises of the REVIZEE Program from 1997 to 1999.

2.2. Deep-water sharks

***Squalus* spp.**: Spiny dogfishes along with the dusky smooth-hound (*M. canis*) are probably the most abundant sharks on the continental slope of Brazil [15, 61, 68, 69]. This genus forms a complex of species where only *Squalus acanthias* Linnaeus 1758 is constantly recorded and recognized. Other species are frequently misidentified and named as *S. mitsukurii-blainvillei* and *S. megalops-cubensis* groups or *Squalus* sp. A and *Squalus* sp. B [18, 24, 70]. Their taxonomic resolution was recently investigated [71] when four new species were described as follows: *Squalus albicaudus, S. bahiensis, S. lobularis,* and *S. quasimodo* Viana, Carvalho and Gomes 2016. According to these authors, *S. albicaudus, S. bahiensis,* and *S. quasimodo* are short-nosed dogfishes, or what was previously grouped as *S. megalops-cubensis*, while *S. lobularis* is a long-nosed dogfish, what was identified as *S. mitsukurii-blainvillei*. Northeastern dogfish specimens were limited in Viana et al. [71] analysis and a precise description on the distribution of these species is urgently needed, however, this is a long waited taxonomic review that brings light on the diversity of the genus in Brazil. Research cruises off the northeastern continental slope in search of deep-water elasmobranchs with bottom longline (15 days cruise and 1000 hooks per line set, each) usually landed 150–250 [or 10–16 ind./1000 h] dogfishes per cruise (probably *S. albicaudus* and *S. lobularis* instead of *S. cubensis-megalops* and *S. mitsukurii* Jordan and Snyder 1903, respectively) (**Figure 3**), a lower CPUE when compared to the southern deep-water gill net fishery which reported a total capture of 12,208 specimens of *S. megalops* (*S. albicaudus*) captured in 14 cruises in 2001 (872 specimens *per* cruise) [15]. Haimovici et al. [1] reported CPUEs of 32–146 ind./1000 hooks with apparent decreasing CPUEs toward the northern extreme of the sampled area (from the States of Rio Grande do Sul to Rio de Janeiro).

Figure 3. Specimen of *Squalus* cf. *albicaudus* captured off the northeastern continental slope along the REVIZEE program.

These large captures suggest that it is likely that these species will become commercially exploited in the near future [45].

Cirrhigaleus asper: The roughskin spurdog (**Figure 4**) was the third most abundant squaloid captured off the northeastern Brazilian coast by the REVIZEE program with about a hundred specimens captured in six cruises (1997–1999) [45]. This species is also reported off south-eastern Brazil (State of São Paulo), but previous results indicate lower abundances than those reported at the northeast coast [2, 33, 52, 72]. Population status is unknown for this species.

***Echinorhinus brucus* (Bonnaterre 1788)**: The bramble shark is frequently reported in low numbers by deep-water bottom longliners along the coast of Brazil [61, 70, 73, 74]. The REVIZEE research cruises off northeast coast usually captured two or three large mature specimens per cruise (15 days cruise/1000 hooks per longline set) (**Figure 4**). This species is also occasionally trawled on the continental shelf off southern Brazil especially during the winter months. The deep-water gill net monitoring program for the monkfish fishery reported an estimation of 4378 specimens captured in 2001 with a CPUE of 0.55 specimens per 100 nets [15]. These by-catch levels bring concern on the conservation status of local populations of the bramble shark since there is restricted biological information on this species [75].

***Centrophorus granulosus* (Bloch and Schneider, 1801)**: This squaliform shark was recorded off the state of Sergipe at 350 m depth where twenty-seven specimens were captured in one single bottom longline set [18, 24, 61] (**Figure 3**). Unfortunately, this was the last line set of the cruise and no other specimen was captured in latter deep-water surveys. The presence of adult males and pregnant females was considered evidence of a local resident population (**Figure 4**). The unique capture suggests a deeper (more than 500 m) bathymetric distribution where scientific surveys with bottom longline were not developed. Although *Centrophorus* is considered to be of commercial importance in many regions [76–78], the Brazilian population is still unknown to the deep-water fishery and this possible stock is considered intact. However, Guallart et al. [79] considered this species particularly vulnerable to fisheries based on its low fecundity rates, long gestation periods, and late maturity.

***Etmopterus* spp.** The Brazilian lanternsharks are currently represented by six species: *E. bigelowi, E. pusillus* (Lowe, 1839), *E. gracilispinis* Krefft 1968, *E. lucifer, E. granulosus* (Günther, 1880), and *E. hillianus* (Poey, 1861) [22, 47, 75, 80–82]. Specific abundances in catches presented a latitudinal shift with *E. bigelowi* (**Figure 4**) being more abundant off the northeast coast, while *E. lucifer* was more abundantly recorded off the south/southeast coast [47, 58, 61, 70, 75, 82]. However, these differences probably resulted from different fishing gears (bottom longline vs. otter trawl). Preliminary stomach content analysis of *E. bigelowi* indicates a benthopelagic feeding habit with predominance, in frequency of occurrence, of crustaceans and cephalopods [83]. Sexual maturity happens at similar sizes for males and females from 60 to 63 cm TL [47]. Perez and Wahrlich [15] and Perez et al. [58] report that lanternsharks are completely discarded onboard.

Apristurus parvipinnis: This shark is occasionally reported from deep-water bottom trawlers (deeper than 600 m) in small numbers, but frequently ignored or misidentified as small squaliforms (*Etmopterus*) due to the dark pigmentation (**Figure 4**). Captures were reported by the REVIZEE Central-Score (off the State of Bahia) and the State of Santa Catarina by commercial fleet; sometimes along with *A. profundorum* (Goode and Bean, 1896), a somewhat more sporadic and geographically restricted species [18, 31, 84, 85].

Galeus mincaronei: This recently described small species is reported only for a restricted area off the State of Santa Catarina at depths of about 200 meters [23, 73, 86]. Its taxonomic resolution still requires a further comparison to *G. antillensis* Springer 1979 [31], but the high level of endemism and vulnerability to deep-water fisheries (bottom longline and otter trawl) brings concern on its occurrence as a common bycatch species [87] (**Figure 4**).

***Parmaturus* cf. *campechiensis*:** This species is a rare and uncommon one in the deep-water elasmobranch assemblage. Its deeper depth range (>600 m) probably protects the local population off southern Brazil from deep-water trawlers. This species was described based on a single immature specimen off Campeche (Mexico) and the Brazilian specimens (female specimens with 410 mm TL and 440 mm TL), if identification is confirmed, probably represent the second record of the species in the world [31, 88] (**Figure 4**). At least three specimens were captured so far, but a formal re-description of the species will be published elsewhere.

Pseudotriakis microdon: One single female specimen (2670 mm TL and 85.5 kg TW) was captured off the State of Paraíba by a bottom longline at 450 m depth (**Figure 3**). The whole animal was discarded after being misidentified as one *Carcharhinus* sp. [17, 18, 24, 61]. Although rare in research cruises, this species may become a bycatch not reported or misidentified in an eventual deep-water longline fishery for *Lopholatilus villari* Miranda-Ribeiro 1915 and *Epinephelus itajara* (Lichtenstein, 1822).

Mitsukurina owstoni: The first Brazilian record was unofficially reported off Rio Grande do Sul by fisherman's descriptions and a presumed photograph of the specimen captured by a deep-water gill net in July of 2001 at 380 m depth, but the animal was discarded on board before scientific analysis. Recent deep-water trawls off the states of Amapá and Pará (north Brazil) by IBAMA as part of a deep-water survey in search of new fishery resources (Proarrasto Project) resulted in the capture of six specimens (2100–2900 mm TL) with only one single preserved

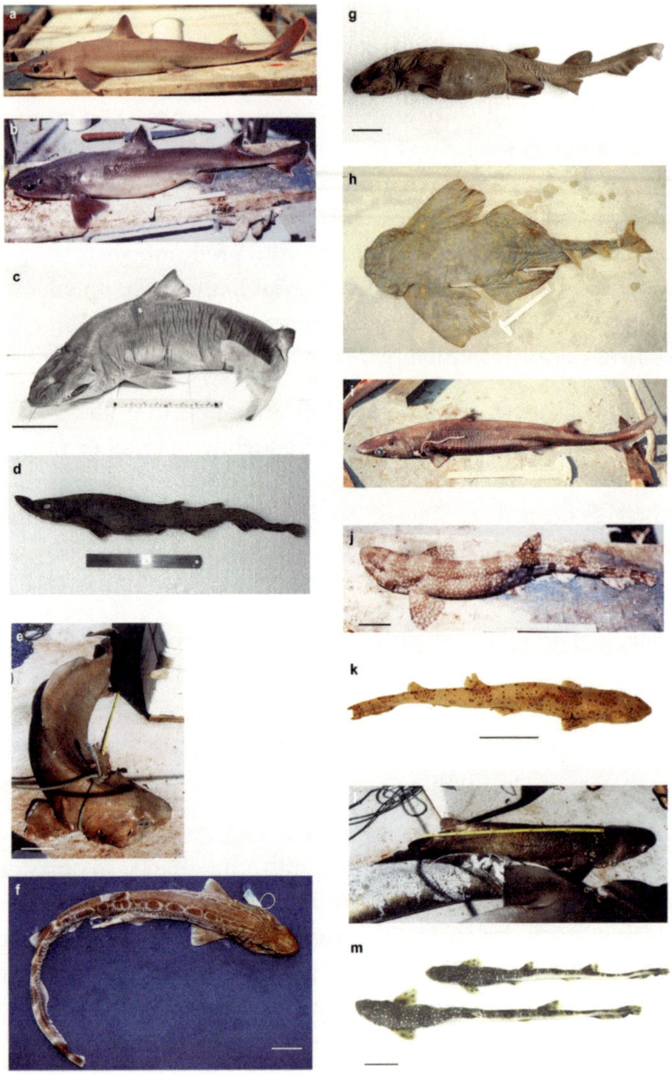

Figure 4. Deep-water sharks and respective scale references: (a) *Squalus* cf. *albicaudus*, 20 mm; (b) *Cirrhigaleus asper*, 150 mm; (c) *Centrophorus granulosus*, 100 mm; (d) *Apristurus parvipinnis* Springer and Heemstra 1979, 150 mm; (e) *Pseudotriakis microdon*, 300 mm; (f) *Galeus mincaronei*, 20 mm; (g) *Parmaturus* cf. *campechiensis*, 40 mm; (h) *Squatina* sp. A, 150 mm; (i) *Etmopterus bigelowi* Shirai and Tachikawa 1993, 150 mm; (j) *Scyliorhinus ugoi*, 40 mm; (k) *Scyliorhinus haeckelii* (Ribeiro, 1907) (holotype MNRJ-494), 60 mm; (l) *Echinorhinus brucus*, 1000 mm; and (m) *Schroederichthys saurisqualus* Soto 2001, 50 mm. All photographs by GR, except the false catshark and bramble sharks, which were photographed by the REVIZEE research team.

specimen (female 1040 mm TL, the smallest of all captured animals) [27, 28, 38]. All other specimens were discarded without further scientific analysis. More recently in 2008, a large mature male (3152 mm TL and 99 kg total weight] was captured in depths of 700–1000 m off the state of Rio de Janeiro by a chartered boat for deep-water shrimp (**Figure 5**) [39].

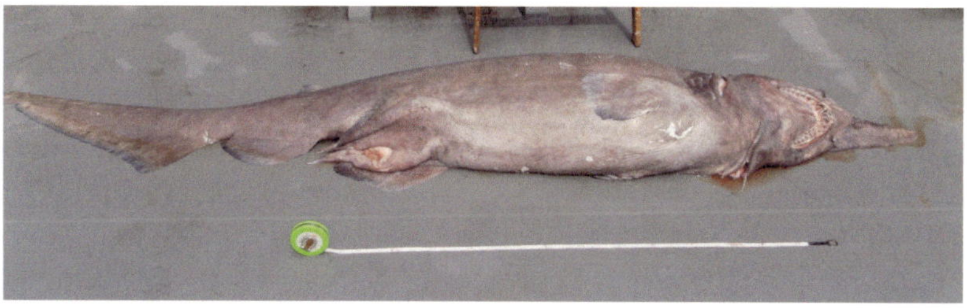

Figure 5. *Mitsukurina owstoni* male specimen with 3152 mm TL captured off Rio de Janeiro in 2008.

Schroederichthys saurisqualus **and** ***S. tenuis***: The recently described catshark *S. saurisqualus* is endemic to southern Brazil with higher captures between the states of Paraná and Rio Grande do Sul. This species presents no economic value and is discarded on board [15]. *S. saurisqualus* is commonly captured by trawlers, longliners and even gill nets in waters deeper than 200 m (**Figure 4**). The slender catshark *S. tenuis* is endemic to the northern coast and has occasional catches on the continental slopes of the states of Amapá and Pará [28, 37, 88]. Both species were captured during the research surveys of the REVIZEE Program and their population impact due to deep-water fishery is currently unknown, but their high endemism level is an additional component of concern.

Scyliorhinus haeckelii: The recent increased interest in the freckled catshark *S. haeckelii* meat has changed its bycatch status to an occasionally commercialized species by trawlers working on the continental shelf off the state of São Paulo. This interest occurs when other more economic species are not present in the captures and catsharks may be processed as fish filets by fishing industries of Santos (southeastern coast). Perez and Wahrlich [15] reported the capture of 4584 specimens in 2001 (estimated CPUE of 0.55 specimens per 100 nets) by the deep-water gill net fishery of southern Brazil. The geographic distribution of the species ranges from Rio Grande do Sul up north to at least Rio de Janeiro. A recent taxonomic review of the *S. haeckelii-besnardi* complex [89] confirms that *S. besnardi* Springer and Sadowsky 1970 is a junior synonym of *S. haeckelii* with a grade on the color pattern [90, 91] (**Figure 4**). A new species (*S. ugoi*) was described on the northeast slope based on specimens with a lighter color pattern on the dorsal surface when still alive [92]. This new species is restricted to upper depths of the continental slope (100–500 m] and the TAMAR Project (Brazilian Marine Turtles Project) succeeded on its captured reproduction. One fourth and larger species is believed to occur on the south continental slope on deep waters (600–1000 m) and is previously identified here as *Scyliorhinus* sp. B.

Squatina **spp**: Three species of angel sharks are recognized along the Brazilian coast: *S. argentina* (Marini, 1930), *S. guggenheim* Marini 1936 and *S. occulta* Vooren and Da Silva 1991 [93]. Two additional forms are known, but both need more analysis and specimens for an appropriate comparison. The first one is recurrently identified as *S. dumeril* Lesueur 1818 (here assigned as *Squatina* sp. A) [17, 18, 94] mostly due to an overall bluish-gray color when preserved, general aspect, similar sizes (970–1200 mm) and replication of superficial identifications rather than considering specific characters and a proper morphometric analysis. However, *Squatina* sp. A is restricted to deep-waters and demands further taxonomic investigation. Both *S. argentina* and *Squatina* sp. A are commonly captured over the continental shelf break and upper continental slope (150–600 m), but only *Squatina* sp. A (**Figure 4**) seems to have larger abundances at deeper waters with captures sporadically reported off the north and northeast coasts of Brazil; probable continuous geographic distribution from Amapá to Espírito Santo (19° S) [17, 18, 37, 59, 94]. Not evaluated by the IUCN Red List, this species may be part of the deep-water bycatch fishery [48]. The fifth putative species *Squatina* sp. B was reported by Vaz and Carvalho [93] as *Squatina* sp., and according to these authors, it differs from all other species on the elevated number of vertebral centra, tooth count, color, position, and size of enlarged dermal denticles. More material and a further analysis are needed before a final position on both forms *Squatina* sp. A and *Squatina* sp. B is assumed.

2.3. Deep-water rays

Gurgesiella dorsalifera: This small skate is commonly captured as bycatch by the southern deep-water squid trawl (**Figure 6**). The species has no economic value due to its small size and is frequently misidentified as *Psammobatis* or other juvenile rajids. Most of the specimens landed were adhered to the squids, but probably a larger proportion was discarded at sea immediately after capture. Its high endemism off southern Brazil is a matter of concern [65]. After considering the endemism and on growing fishery risk, it was listed as a vulnerable species by the IUCN Red List of Threatened Species [58, 64, 67, 95–97].

Benthobatis kreffti: This recently described small species of blind electric ray is reported only off south/southeastern Brazil where it is occasionally found as bycatch in the deep-water otter trawl squid fishery [25, 97, 98] (**Figure 6**). Its small size, glandular shape and absolute no economic value makes it easier to be ignored by onboard observers among the trawled squid. Due to its presumed feeding aggregation over polychaete-rich sediments, one single hour trawl may capture a large number that perfectly fits in a small plastic basket. Commercial and scientific cruises were able to capture more than 150 specimens in a couple of trawls, which added to the low fecundity of the species (maximum of 2 embryos per gestation period) [98] makes it a vulnerable animal to any fishery effort.

***Dipturus* spp**: Four species of *Dipturus* are reported for the Brazilian continental slope and tree species are under analysis: *D. teevani* (Bigelow and Schroeder, 1951), *D. leptocaudus* Krefft and Stehmann 1974, *D. mennii* Gomes and Paragó 2001, *D. bullisi* (Bigelow and Schroeder, 1962), *Dipturus* cf. *garricki* (Bigelow and Schroeder, 1958) and *Dipturus* sp. A and *Dipturus* sp. B [17, 19, 26, 36, 64, 99, 100]. However, a taxonomic review of this genus is strongly necessary since several misidentifications and descriptions seem to have occurred. The genus *Dipturus* comprises a complex and morphologically conservative group with diagnostic characters that change during the animal growth and maturity [101]. The ontogenetic morphological variation makes it difficult for the specific identification of neonates and juveniles by onboard observers and specimens are frequently misidentified as *R. agassizi* or *Sympterygia acuta* Garman 1877. *Dipturus trachydermus* Krefft and Stehmann 1974 is regularly recorded, but its occurrence needs confirmation [15, 102]. Large specimens are commercialized, but a great number of neonates and juveniles are discarded as bycatch (**Figure 6**). Newly hatched

http://dx.doi.org/10.5772/intechopen.69471

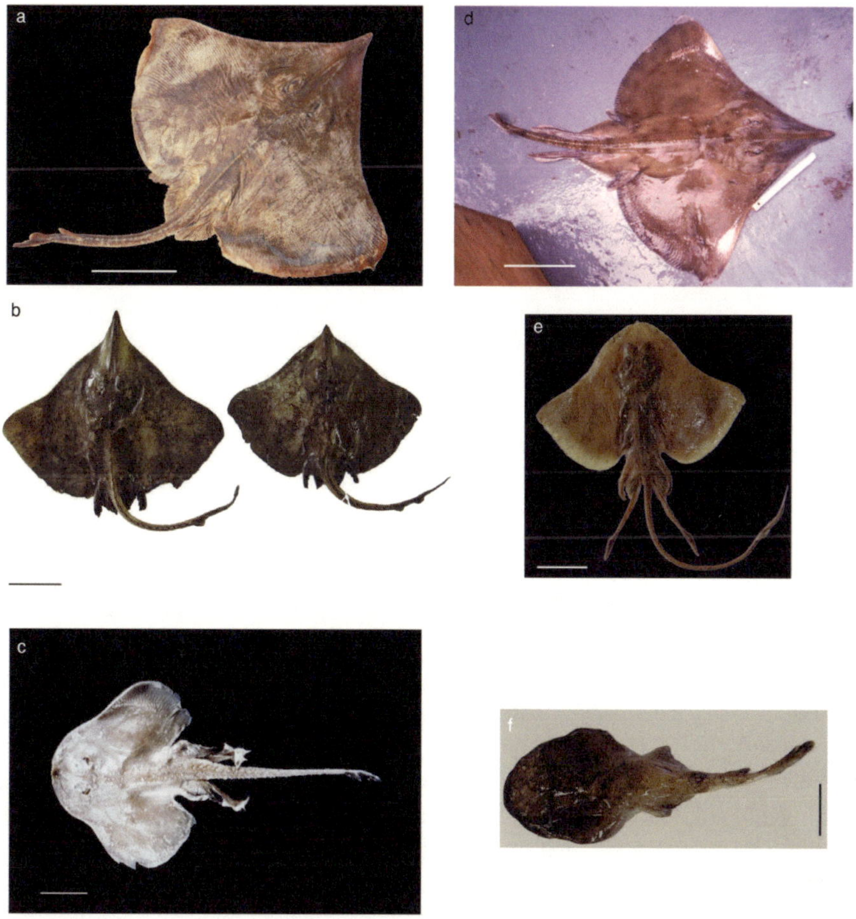

Figure 6. Deep-water rays and respective scale references: (a) *Dipturus mennii*, 150 mm; (b) *Dipturus* spp. neonates, 20 mm; (c) *Breviraja spinosa*, 50 mm; (d) *Dipturus* sp. B, 300 mm; (e) *Gurgesiella dorsalifera*, 30 mm; (f) *Benthobatis kreffti*, 25 mm.

specimens of *Dipturus* spp. with internal yolk reserves were captured by the deep-water otter trawl squid fishery in late winter (August–September) off southern Brazil, which brings concern about the future of these populations. The gill net monkfish fishery at the same area also captured mature specimens of *Dipturus* spp. (*D. trachydermus* and *D. mennii*) and these rays were the most abundant elasmobranch bycatch species with a total annual capture of 23,954 specimens and a CPUE of 2.89 ind./100 nets [15]. Visintin and Perez [102] still identify *Dipturus* as an extremely vulnerable group of species to the gill net fishery based on productivity-susceptibility analysis (PSA).

***Tetronarce puelcha* Lahille 1928**: The Argentine torpedo ray (*Tetronarce puelcha*), supposedly rare and never abundant in scientific and commercial captures off Brazil, Uruguay, and Argentina [103–105], was the third most abundant bycatch species captured by the deep-water gill net fishery off southern Brazil (estimated catch of 10,391 specimens captured in 2001 based on an estimated CPUE and *Tetronarce*/monkfish ratio) [15]. These captures corroborate the hypothesis that *T. puelcha* is a member of the deep-water chondrofauna instead of a neritic species [106]. The northern Atlantic torpedo ray (*T. nobiliana*) was reported at deep-waters off the states of Amapá and Pará by scientific surveys of the REVIZEE program (otter trawl), and states of São Paulo and Rio de Janeiro [58], but due to the singularity of these records and lack of preserved specimens, this identification was not herein considered definitive.

Other rare deep-water species were recorded along the Brazilian coast, but most of them are known by single records or sporadic captures, e.g., *S. antarcticus* Whitley 1939, *H. griseus, B. spinosa, B. nigriventralis* McEachran and Matheson 1995, *D. leptocaudus, D. clarkii* (Bigelow and Schroeder 1958), *Malacoraja obscura* Carvalho, Gomes and Gadig 2005, *Diplobatis pictus* Palmer 1950 and *C. rugosa* Bigelow and Schroeder 1958 [15, 30, 36, 37, 58, 61, 75, 85, 107–110]. These species probably are distributed in deeper waters and are currently inaccessible to the Brazilian deep-water fishery. However, due to the large discard of unknown or unwanted small species in these fisheries, an accurate estimation of their contribution is presently speculative.

2.4. Chimaeras

Chimaeriforms (Holocephali: Chimaeriformes) are known in Brazil as "quimeras" and this designation includes rabbit fish, elephant fish, longnose chimaera, and all other species of this group. Four recognized species, one putative species and one unidentified species are recorded as follows: *Callorhinchus callorynchus* (Linnaeus, 1758), *Hydrolagus alberti* (Bigelow and Schroeder, 1951), *Hydrolagus matallanasi* (Soto and Vooren, 2004), *Harriotta raleighana* (Goode and Bean, 1895), *Hydrolagus* cf. *mirabilis* (Collett, 1904) and *Rhinochimaera* sp. The American Elephantfish *C. callorynchus* was the first chimaera identified and included in Figueiredo's [105] catalog of species of the state of São Paulo. This species is occasionally recorded along the continental shelf during winter months from the states of Rio Grande do Sul to São Paulo always in small numbers and captured by otter trawl or gill net. All other records are typical deep-water species, but only *H. matallanasi* [29] is frequently captured and reported on REVIZEE reports [52, 97], and bycatch monitoring programs [15, 58] (**Figure 7**).

Figure 7. Specimens of *H. matallanasi* captured by commercial fleet on the continental slope of Santa Catarina.

All other records are based on single or restricted captures [36, 111]. The lack of information on the group brings concern on its sustainability under a fishing pressure where it composes the bycatch.

3. Discussion

The concept of bycatch is somewhat controversial (see [112] for a review) and intangible for deep-water fisheries as captures may vary drastically (spasmodically) from different fishery grounds, over time and in percentage of species composition [4, 7]. In addition, market interests frequently make fishing effort to shift quickly from one species to another of previously low economic value or even completely ignored. An appropriate approach to the bycatch problem must meet the fisheries' management needs providing means for a realistic evaluation of the fishery impact on at least all populations of captured species when indirect effects of the fishery cannot be estimated. This means a specific identification of all capture including discarded and landed species [77, 78, 113] in order to provide managers with reliable information for a complete assessment. This requires well trained onboard observers and a complete capture description, not only the economically relevant portion. These observers have the tendency to give more attention to abundant, larger, and well known species, ignoring small unfamiliar creatures with occasional occurrences. However, this part of the capture is equally susceptible to easy overfishing as they are under the very same extreme environmental and ecological constrains of the deep-water habitats ([114, 115] for a review). Hence, catch levels of target species should be estimated based on the most fragile population compounding the capture—the weak link of the chain—regardless of it being a target or bycatch species. Generic categorizations should be avoided [10, 78] and a complete field guide of all captured species with an identification key would be expected to be published and available for the onboard observers as soon as the first cruises are monitored. Those species without a proper description should be treated at the species level as A, B, C, and so on (*Squalus* sp. A, e.g., [116]), and new records saved on board for posterior laboratorial identification by taxonomists [58].

The monkfish fishery case off southern Brazil is a perfect example of an ongoing well reported and studied deep-water fishery [4, 8, 15, 102, 117, 118]. However, unfortunately the IBAMA did not require any species-catch limit (quota) permission or established a precautionary approach prior to its development [4, 8]. All actions related to the management of the fishery came considerably after the catch/landing have already started and signed for a near future effort increase [118]. These actions aimed to collect fishery and biological information in order to elaborate a management plan for the monkfish fishery [8] and to avoid conflicts with the national fleet. According to Perez et al. [118], a total biomass reduction of 16 to 60% (more realistic) occurred in the main fishing grounds in 1 year (2001). As a result, Perez et al. [8] proposed a series of management actions to immediately reduce fish mortality to precautionary levels and a total allowable catch of 2500 mt. However, all management actions were developed based solely on the catch levels and population dynamics of the monkfish, not considering the biological characteristics or catch levels of the bycatch species, especially those of elasmobranchs. Since deep-water elasmobranchs seem to present an extreme k-strategy and high extinction risk [16, 46] and may reach longevities of 70 years and age of first maturity of 44 years (Age_{50}/Age_{max} = 0.62) (*Centrophorus squamosus* (Bonnaterre, 1788), [116, 119]), it is reasonable to assume that elasmobranchs are one of "the weak links of the chain" in deep-water fisheries and any management action must consider their most fragility to fishery effort. Otherwise, some bycatch deep-water elasmobranchs are under the risk of unnoticed population collapses or even local extinctions [120, 121] before target species presents any sign of exhaustion.

The Brazilian deep-water elasmobranch diversity is poorly known and the recent increase in species records and descriptions are evidences of a possible unevaluated richness along the continental slope [17, 31, 35, 36, 75]. Large species such as *M. owstoni, P. microdon, S. antarcticus,* and *C. granulosus* were only reported when scientific efforts in search of deep-water resources were recently developed (PROARRASTO and REVIZEE programs) [27, 37, 38, 61, 83]. Therefore, if a larger effort targeting deep-water elasmobranchs is performed, it would probably change the beliefs about the rarity of these animals [31, 39]. The number of reported species drastically increased from 119 to 176 along the past decade (**Figure 1**), which corresponds to a total number increase of 36%, or 63% in numeric terms of skates and rays and 16% of sharks [17, 18, 36]. This recent biological diversity boom is certainly related to the REVIZEE scientific surveys in the recent past (1996–2007) and its ongoing influence, which resulted in the most extensive effort in order to evaluate possible promising marine fishery resources in Brazil [52, 75, 122]. However, even these scientific surveys need to be properly designed to catch and better evaluate the elasmobranch biodiversity. The restricted size and operational capability of most Brazilian research vessels usually demand a small and multitasked scientific crew, able to catch and process a wide range of organisms from cephalopods to elasmobranchs, and from physics oceanographic data collection to plankton net sets. This, allied to taxonomic identification problems, lack of information on the rarity records of some species, large size of sharks and onboard deterioration due to poor ice conditioning may be responsible for previously identified problems such as onboard discards, inappropriate deposition of rare specimens in institutions without proper scientific report and biological information loss [17, 37, 61].

Deep-water species are commonly treated as the last possible fishery resources—the dead end of fisheries—with unexpected sustainability and intrinsic need of a precautionary approach and last refuge for coastal species [10, 16]. Gordon [7, 78] proposed principles and actions based on the North Atlantic deep-water fishery experience in order to create means for a possible fishery management. Perez et al. [8, 118], Perez and Wahrlich [15], and Visintin and Perez [102] have proposed a series of management actions toward the monkfish fishery off southern Brazil. These proposals are related mainly to fishery area restrictions or closures, bycatch reduction and fleet-effort control. Some additional aspects are proposed below considering species diversity in the management of deep-water fisheries based on all results herein discussed:

1. Specific bycatch identification is primordial for any management program. The use of fish categories should be completely banned in deep-water fisheries.

2. All catch must be monitored onboard, especially discarding bycatch of small species, which may have an extreme k-strategy like large species but are frequently captured in large numbers without calling attention to its population decline. The same principle is applicable to discarded or retained rare species, for which the point of economic extinction—death is inevitable for discarded or retained species—may be closer to the point of biological extinction because it is still economical to continue capturing rare bycatch species as long as the target species is still viable [123]. Therefore, the population size at which economic extinction occurs will move down closer to biological extinction than in the case of target species [121].

3. Onboard observers programs must be developed and/or improved in a larger proportion of the deep-water fleet along with means for reliable species identification (species catalogues, training, methodology, and scientific personnel). Scientific surveys are necessary in order to compare catch efficiency between different fishing gears and revision of methodologies and taxonomic approach (identification guides) of the onboard observers program.

4. Since the great majority of deep-water species are extreme k-strategists [116], fisheries management measures should be based on the most fragile (weak links), endemic, or most vulnerable species of the catch.

5. Fishing gears should be evaluated in selectivity terms and specific composition of the catch, in particular fishing grounds due to the contagious distribution of some deep-water species. Species richness and relative abundances may vary significantly as a direct interaction between benthic species and the sea floor type.

6. The assumption of the precautionary principle in fisheries requires a complete effort and fleet control, which means a restricted and monitored number of boats permitted to fish deep-water resources in specific areas and fishing grounds. Deep-water fisheries should not receive subsidies allowing fast growth without a proper scientific side-by-side development.

7. Deep-water resources should not be seen as economical alternatives to continental shelf collapsed resources; instead, the deep-water habitats should be considered as the new candidates for conservation [10]. The "gold rush" behavior must be controlled in order to successfully implement the precautionary principle (catch rates, effort control, monitoring and onboard programs, fishing areas, and others).

8. The proportion of endemic species in the catch should be considered when deep-water fisheries are being evaluated for implementation or management. Deep-water sharks and rays are often assumed as wide geographical range species, but when a relevant proportion (>5%) of deep-water endemic species makes part of the elasmobranch assemblage—catch, an additional precautionary reinforcement should be added.

These proposals and arguments for the Brazilian deep-water fishery management were not finally stressed and as new regions are considered to deep-water longline, trawl or gillnetting, particular and occasional needs will ask different approaches and restrictions. How hard the species diversity component will push on these management decisions is our deepest concern.

Acknowledgements

The authors would like to express their gratitude to the REVIZEE Program, Scores North, Northeast and South; DIMAR and LOP laboratories, students and researchers at the UFRPE; and Drs. Carolus Maria Vooren and Miguel Petrere Júnior for their guidance and help. The senior author was a REVIZEE and PhD granted from CNPq and CAPES Brazilian agencies.

Author details

Getulio Rincon[1]*, Rodrigo Cordeiro Mazzoleni[2], Ana Rita Onodera Palmeira[3] and Rosangela Lessa[4]

*Address all correspondence to: getulio.rincon@ufma.br

1 Federal University of Maranhão—UFMA, Center of Human, Nature, Health and Technology Sciences—CCHNST, Pinheiro-Maranhão, Brazil

2 University of the Vale do Itajaí—UNIVALI, Center of Earth and Ocean Technological Sciences—CTTMar, Santa Catarina, Brazil

3 Federal University of Paraíba—UFPB, Center of Exact and Natural Sciences, University City, João Pessoa, Paraíba, Brazil

4 Federal Rural University of Pernambuco—UFRPE, Department of Fisheries Engeneering, Recife-PE, Brazil

References

[1] Haimovici M, Silva AOA, Tutui SLS, Bastos GCC. Prospecção pesqueira demersal – Espinhel de fundo. Relatório Técnico de Atividades 1997. Rio Grande-FURG: Programa REVIZEE Score Sul; 1998. p. 58

[2] Tutui SLS, Bastos GCC, Tomás ARG, Tiago GG, Zavala-Camin LA. Species composition of the exploratory fisheries with bottom longline off southeastern Brazil. Ciencia e Cultura. 2000;**52**(1):55-58

[3] Perez JAA, Pezzuto PR, Andrade HA. Biomass assessment of the monkfish *Lophius gastrophysus* stock exploited by new deep-water fishery in southern Brazil. Fishery Research. 2005;**72**:149-162. DOI: 10.1016/j.fishres.2004.11.004

[4] Perez JAA, Wahrlich R, Pezzuto PR, Schwingel PR, Lopes FRA, Rodrigues-Ribeiro M. Deep-sea fishery off southern Brazil: Recent trends of the Brazilian fishing industry. Journal of Northwest Atlantic Fishery Science. 2003;**31**:1-18

[5] Perez JAA, Wahrlich R, Pezzuto PR. Chartered trawling on the slope off Brazilian coast. Marine Fisheries Review. 2009;**71**(2):24-36

[6] Kiraly SJ, Moore JA, Jasinski PH. Deepwater and other sharks of the U.S. Atlantic Ocean Exclusive Economic Zone. Marine Fisheries Review. 2003;**65**(4):1-20

[7] Gordon JDM. Deep-water fish and fisheries: Introduction. Fishery Research. 2001;**51**:105-111. DOI: 10.1016/S0165-7836(01)00238-7

[8] Perez JAA, Wahrlich R, Pezzuto PR, Lopes FRA. Estrutura e dinâmica da pescaria do peixe-sapo *Lophius gastrophysus* no sudeste e sul do Brasil. Boletim do Instituto de Pesca. 2002;**28**(2):205-231

[9] Pezzuto PR, Borzone CA. The scallop *Pecten ziczac* [Linnaeus, 1758] fishery in Brazil. Journal of Shellfish Research. 1997;**16**(2):527-532

[10] Morato T, Watson R, Pitcher TJ, Pauly D. Fishing down the deep. Fish and Fish. 2006;**7**:24-34. DOI: 10.1111/j.1467-2979.2006.00205.x

[11] Branch TA, Hilborn R, Haynie AC, Fay G, Flynn L, Griffiths J, Marshall KN, Randall JK, Scheuerell JM, Ward EJ, Young M. Fleet dynamics and fishermen behavior: Lessons for fisheries managers. Canadian Journal of Fisheries and Aquatic Sciences. 2006;**63**(7):1647-1668. DOI: 10.1139/f06-072

[12] Jablonski S. Mar-oceanografia/biologia pesqueira. Parcerias. Estratégicas.2005;**20**(pt. 2): 857-901

[13] Castello HP. Gestão sustentável dos recursos pesqueiros, isto é realmente possível? Pan-American Journal of Aquatic Sciences. 2007;**2**(1):47-52

[14] Vaske T. Seabirds mortality on longline fishing for tuna in southern Brazil. Ciencia e Cultura. 1991;**43**:388-390

[15] Perez JAA, Wahrlich R. A by-catch assessment of the gillnet monkfish *Lophius gastrophysus* fishery off southern Brazil. Fishery Research. 2005;**72**:81-95. DOI: 10.1016/j.fishres.2004.10.011

[16] García VB, Lucifora LO, Myers RA. [2007] The importance of habitat and life history to extinction risk in sharks, skates, rays and chimaeras. Proceedings of the Royal Society B. 2008;**275**(1630):83-89. DOI: 10.1098/rspb.2007.1295

[17] Lessa R, Santana FM, Rincon G, Gadig OB, El-Deir ACA. Biodiversidade de elasmobrânquios do Brasil. Relatório Nécton-Elasmobrânquios. Projeto de Conservação e Utilização Sustentável da Diversidade Biológica Brasileira [PROBIO] [Internet]. Ministério do Meio Ambiente [MMA]. 1999.p. 154. http://www.brasil-rounds.gov.br/round5/arquivos_r5/guias/perfuracao/5round/refere/Elasmobranquios.pdf [Accessed: 2017-03-10]

[18] Gadig OGF. Tubarões da costa brasileira. [thesis]. Rio Claro, São Paulo: Instituto de Biociências de Rio Claro, Programa de Pós-Graduação em Ciências Biológicas-Zoologia. Universidade Estadual Paulista-UNESP; 2001

[19] Gomes UL, Paragó C. Espécie nova de Rajídeo [Chondrichthyes, Rajiformes] do Atlântico sul ocidental. Boletim do Museu Nacional. 2001;**448**:1-10

[20] Soto JMR. First record of kitefin shark, *Dalatias licha* [Bonnaterre, 1788] [Chondrichthyes, Dalatiidae], in the South Atlantic. Mare Magnum. 2001;**1**(1):23-26

[21] Soto JMR. *Schroederichthys saurisqualus* sp. nov. [Carcharhiniformes, Scyliorhinidae], a new species of catshark from Southern Brazil, with further data on *Schroederichthys* species. Mare Magnum. 2001;**1**(1):37-50

[22] Soto JMR. First record of southern lanternshark, *Etmopterus granulosus* [Günther, 1880] [Squaliformes, Dalatiidae], in Brazilian coast. Mare Magnum. 2001;**1**(1):7-10

[23] Soto JMR. *Galeus mincaronei* sp. nov. [Carcharhiniformes, Scyliorhinidae] a new species of sawtail catshark from southern Brazil. Mare Magnum. 2001;**1**(1):11-18

[24] Soto JMR. Annotated systematic checklist and bibliography of the coastal and oceanic fauna of Brazil. I Sharks. Mare Magnum. 2001;**1**(1):51-120

[25] Rincon G, Stehmann MFW, Vooren CM. Results of the research cruises of FRV 'Walther Herwig' to South America. LXXIV. *Benthobatis kreffti* n. sp. [Chondrichthyes, Torpediniformes, Narcinidae], a new deep-water electric ray from off South Brazil and the third species of the genus. Archive of Fishery and Marine Research. 2001;**49**(1):45-60

[26] Gomes UL, Costa SR. New records of the thintail skates *Dipturus leptocauda* (Rajidae, Rajinae, Rajini), with notes on its taxonomy. Biociências. 2003;**11**:91-95

[27] Soto JMR, Asano-Filho M, Holanda FCAF. Sobre a presença de tubarão-duende *Mitsukurina owstoni* Jordan, 1898 [lamniformes, Mitsukurinidae] na costa brasileira. In: Proceedings of the IV Reunião da Sociedade Brasileira para o Estudo de Elasmobrânquios-SBEEL;November 2004; Recife-PE, Brazil. p. 180

[28] Soto JMR, Mincarone MM, Asano-Filho M, Souza RFC. Novos registros de elasmobrânquios coletados em águas profundas ao largo da costa do Amapá. In: Proceedings of the V Reunião da Sociedade Brasileira para o Estudo de Elasmobrânquios-SBEEL;November 2006; Itajaí-SC, Brazil. p. 81

[29] Soto JMR, Vooren CM. *Hidrolagus matallanasi* sp. nov. [Holocephali, Chimaeridae] a new species of rabitfish from southern Brazil. Zootaxa. 2004;**687**:1-10

[30] Carvalho MR, Gomes UL, Gadig OBF. Description of a new species of skate of the genus *Malacoraja* Stehmann, 1970: The first species of the southwestern Atlantic Ocean, with notes on generic monophyly and composition [Chondrichthyes, Rajidae]. Neotropical Ichthyology. 2005;**3**(2):239-258. DOI: 10.1590/S1679-62252005000200002

[31] Rincon G, Vooren CM. Taxonomic and biological records on the south Atlantic marbled catshark, *Galeus mincaronei* Soto [Elasmobranchii:Scyliorhinidae], off the Southern Brazilian Coast. Pan-American Journal of Aquatic Sciences. 2006;**1**(1):1-7

[32] Bezerra MA, Furtado-Neto MAA, Gadig OBF. Fauna de tubarões da costa norte e nordeste do Brasil – do Rio Grande do Norte ao Amapá. In: Proceedings of the V Reunião do Grupo de Trabalho sobre Pesca e Pesquisa de Tubarões e Raias no Brasil;November 1991. Santos-SP, Brazil. p. 6

[33] Amorim AF, Gadig OBF, Arfelli CA. Sharks occurrence in the Brazilian coast.In: Proceedings of the VII Reunião do Grupo de Trabalho sobre Pesca e Pesquisa de Tubarões e Raias no Brasil;November 1994; Rio Grande-RS, Brazil. p. 19

[34] Rincon G, Vooren CM, Stehmann M. List of species of skates and rays off Brasil.An update based on captures and bibliography. In: Proceedings of the I Reunião da Sociedade Brasileira para o Estudo de Elasmobrânquios-SBEEL. November 1997. Salvador, BA. p. 135

[35] Soto JMR. Fauna de elasmobrânquios do Brasil: De inexpressiva à quarta mais diversificada do mundo em apenas 20 anos. In: Proceedings of the V Reunião da Sociedade Brasileira para o Estudo de Elasmobrânquios-SBEEL. November 2006. Itajaí-SC, Brazil. p. 62

[36] Rosa RS, Gadig OBF. Conhecimento da diversidade dos Chondrichthyes marinhos no Brasil: A Construção de José Lima de Figueiredo. Arquivos de Zoologia. 2014;**45**(esp.):89-104. DOI: 10.11606/issn.2176-7793.v45iespp89-104

[37] Rincon G. Relatório Parcial das atividades sobre elasmobrânquios [Belém-PA] provenientes dos cruzeiros de prospecção do programa REVIZEE, Score-NO. UFRPE, Recife-PE. Programa REVIZEE Score-NE. 1999. p. 6

[38] Asano-Filho M, Holanda FCAF, Santos FJS, Cavalcante-Júnior TS. Recursos pesqueiros de grandes profundidades na costa norte do Brasil. Instituto Brasileiro do Meio Ambiente e dos Recursos Naturais Renováveis. Diretoria de Fauna e Recursos Pesqueiros. Cepnor 2005. Brasília, IBAMA-MMA. p. 82

[39] Rincon G, Vaske-Júnior T, Gadig OBF. Record of the goblin shark *Mitsukurina owstoni* [Chondrichthyes: Mitsukurinidae] from the south-western Atlantic. Marine Biodiversity Records. 2012;**5**:1-5. DOI: 10.1017/S1755267211000923

[40] Ramos-Porto M, Silva KCA, Viana GFS, Cintra IHA. Camarões de profundidade coletados no Norte do Brasil [Crustacea: Penaeidea e Caridea]. Trabalhos do Instituto de Oceanografia da Universidade Federal de Pernambuco. 2000;**28**(1):71-85

[41] Dallagnolo R, Perez JAA, Pezzuto PR. O ordenamento da pescaria de camarões-de-profundidade [Decapoda: Aristeidae] no sudeste e sul do Brasil. Brazilian Journal Aquatic Science Technology. 2009;**13**(2):31-43. DOI:10.14210/bjast.v13n2.p31-43

[42] Pezzuto PR. Avaliação dos Camarões de profundidade (Decapoda: Aristeidae). In: Pinheiro M, Boos H, editors. Livro Vermelho dos Crustáceos do Brasil: Avaliação 2010-2014. Porto Alegre, RS: Sociedade Brasileira de Carcinologia – SBC; 2016.pp. 82-92.Cap 4

[43] UNIVALI/CTTMAR/GEP. Boletim estatístico da pesca industrial de Santa Catarina – Ano 2002. Itajaí, SC. [Internet]. 2003. Available from: http://gep.acad.univali.br/download/pdf/boletim_2002.pdf [Accessed: 2017-03-10]

[44] UNIVALI/CTTMAR/GEP Boletim estatístico da pesca industrial de Santa Catarina – Ano 2003. Itajaí, SC. [Internet]. 2004. Available from: http://gep.acad.univali.br/download/pdf/boletim_2003.pdf [Accessed: 2017-03-10]

[45] Fischer AF, Veras DP, Hazin FHV, Broadhurst MK, Burgess GH, Oliveira PGV. Maturation of *Squalus mitsukurii* and *Cirrhigaleus asper* [Squalidae, Squaliformes] in the southwestern equatorial Atlantic Ocean. Journal of Applied Ichthyology. 2006;**22**(6):495-501. DOI: 10.1111/j.1439-0426.2006.00749.x

[46] Kyne PM, Simpfendorfer CA. A Collation and Summarization of Available Data on Deepwater Chondrichthyans: Biodiversity, Life History and Fisheries.Seattle-WA, USA: IUCN SSC Shark Specialist Group Report for the Marine Conservation Biology Institute; p. 137

[47] Gianeti MD. Distribuição especial e composição das populações do de tubarões do gênero *Etmopterus* do talude continental do sul do Brasil [thesis]. Rio Grande-RS, Brazil. Instituto de Oceanografia, Fundação Universidade de Rio Grande-FURG; 2005

[48] Compagno LJV, Dando M, Fowler S. Sharks of the World. Princeton Field Guides. Princeton, New Jersey: Princeton University Press; 2005. p. 368

[49] Priede IG, Froese R, Bailey DM, Bergstad OA, Collins MA, Dyb JE, Henriques C, Jones EG, King N. The absence of sharks from abyssal regions of the world's oceans. Proceedings of the Royal Society B. 2005;**273**(1592):1435-1441. DOI: 10.1098/rspb.2005.3461

[50] Ebert DA, Fowler S, Compagno L. Sharks of the World: A fully Illustrated Guide. Plymouth: Wild Nature Press; 2013. p. 528

[51] Last PR, White WT, Carvalho MR, Séret B, Stehmann MFW, Naylor GJP, editors. Rays of the World. Ithaca and London: Cornel University Press; 2016. p. 790

[52] Haimovici M, Avila-da-Silva AO, Rossi-Wongtschowski CLDB, editors. Prospecção pesqueira de espécies demersais com espinhel-de-fundo, na região Sudeste-Sul do Brasil. Série Documentos REVIZEE– Score Sul. São Paulo: Instituto Oceanográfico – USP; 2004. p. 112

[53] Holanda FCAF. Desenvolvimento tecnológico, por métodos de arrasto de fundo, na captura de recursos demersais de profundidade da zona econômica exclusive (ZEE) norte do Brasil [thesis]. Fortaleza-CE: Universidade Federal do Ceará, Programa de Pós-Graduação em ciências Marinhas Tropicais; 2012

[54] Fagundes-Netto EB, Gaelzer LR, Carvalho WF, Costa PAS. Prospecção de recursos demersais com armadilhas e pargueiras na região central da Zona Econômica Exclusiva entre Salvador-BA (13°S) e o Cabo de São Tomé-RJ (22°S). In: Costa PAS, Martins AS, Olavo G, editors. Pesca e potenciais de exploração de recursos vivos na região central da Zona Econômica Exclusiva brasileira. Rio de Janeiro: Museu Nacional; 2005. pp. 129-143

[55] Martins AS, Olavo G, Costa PAS. Recursos demersais capturados com espinhel de fundo no talude superior da região entre Salvador (BA) e o Cabo de São Tomé (RJ). In: Costa PAS, Martins AS, Olavo G, editors. Pesca e potenciais de exploração de recursos vivos na região central da Zona Econômica Exclusiva brasileira. Rio de Janeiro: Museu Nacional; 2005. pp. 109-128

[56] Perez JAA, Wahrlich RA. Bycatch assessment of the gillnet monkfish *Lophius gastrophysus* fishery off southern Brazil. Fisheries Research. 2005;**72**:81-95. DOI: 10.1016/j.fishres.2004.10.011

[57] UNIVALI/CTTMAR/GEP. Boletim estatístico da pesca industrial de Santa Catarina – Ano 2001. Itajaí, SC. [Internet]. 2002. Available from: http://gep.acad.univali.br/download/pdf/boletim_2001.pdf [Accessed: 2017-03-10]

[58] Perez JAA, Pereira BN, Pereira DA, Schroeder R. Composition and diversity patterns of megafauna discards in the deep-water shrimp trawl fishery off Brazil. Journal of Fish Biology. 2013;**83**:804-825. DOI: 10.1111/jfb.12141

[59] Vooren CM. Demersal elasmobranchs. In: Seeliger U, Odebrecht C, Castello JP, editors. Subtropical Convergence Environments: The coast and sea in the Southwestern Atlantic. Heidelberg: Springer-Verlag; 1997. pp. 141-146

[60] Perez JAA, Pezzuto PR, Wahrlich R, Soares ALS. Deep-water fisheries in Brazil: History status and perspectives. Latin American Journal of Aquatic Resources. 2009;**37**(3):513-542. DOI:10.3856/vol37-issue3-fulltext-18

[61] Rincon G. Tubarões e Raias do Talude do Nordeste, Idade e Crescimento de *Squalus mitsukurii*, Análise Morfométrica de *Pseudocarcharias kamoharai*, Desenvolvimento Embrionário de *Prionace glauca*, Análise do Desembarque de Barbatanas pela Frota Pesqueira de Natal-RN. Recife-PE: Relatório Final, Programa REVIZEE Score-NE; 1999. p. 81

[62] Oliveira VS. Distribuição e abundância relativa de peixes e crustáceos capturados no programa REVIZEE/SCORE-NE na plataforma externa e talude da costa do Nordeste do Brasil [thesis]. Recife-PE: Universidade Federal de Pernambuco-UFPE, Programa de Pós-Graduação em Oceanografia; 2005

[63] Vaske T, Rincon G, Vooren CM. Notes on feeding and reproduction of the ray *Gurgesiella dorsalifera* [Elasmobranchii:Rajidae] from southern Brazilian coast. In: Proceedings of the I Reunião da Sociedade Brasileira para o Estudo de Elasmobrânquios-SBEEL;November 1997. Salvador-BA. p. 144

[64] Menni RC, Stehmann MFW. Distribution, environment and biology of batoid fishes off Argentina, Uruguay and Brasil. A review. Revista del Museo Argentino de Ciencias Na turales. 2000;**2**(1):69-109

[65] Rincon G, Vaske-Júnior T, Vooren CM. Stomach contents and notes on the reproduction of the onefin skate *Gurgesiella dorsalifera* (Chondrichthyes: Rajidae) off southern Brazil. Neotropical Ichthyology. 2008;**6**(4):689-692. DOI: 10.1590/S1679-62252008000400019

[66] Rincon G. *Benthobatis kreffti* [Internet].2006. In: IUCN Red List of Threatened Species. Available from: www.iucnredlist.org [Accessed: 2017-03-10]

[67] Rincon G. *Gurgesiella dorsalifera* [Internet]. 2006. In: IUCN Red List of Threatened Species. Available from: www.iucnredlist.org [Accessed: 2017-03-10]

[68] Souto CFM. Estudo comparativo de reprodução nos cações *Mustelus schmitti* Springer, 1839 e *M. canis* Mitchill, 1815 [Pisces: Squaliformes], na plataforma continental do Rio Grande do Sul, Brasil [thesis]. Rio Grande-RS, Brazil: Universidade Federal de Rio Grande, Instituto de Oceanografia; 1986

[69] Calderón JJD. Morfologia, distribuição, abundância e reprodução de duas espécies do gênero *Squalus* no sul do Brasil [thesis]. Rio Grande-RS, Brazil: Universidade Federal de Rio Grande, Instituto de Oceanografia; 1994

[70] Mazzoleni RC, Schwingel PR. Elasmobranch species landed in Itajaí harbor, southern Brazil. Notas Téc Facimar. 1999;**3**:111-118

[71] Viana STF, Carvalho MR, Gomes UL. Taxonomy and morphology of species of the genus *Squalus* Linnaeus, 1758 from the Southwestern Atlantic Ocean (Chondrichthyes: Squaliformes:Squalidae). Zootaxa. 2016;**4133**(1):001-089

[72] Gadig OBF. Taxonomia e sistemática de Elasmobrânquios. Relatório Técnico de Atividades 1997. Programa REVIZEE Score-Sul Report. 1998. p. 11

[73] Soto JMR, Nisa W, Mazzoleni RC. Novos registros de *Echinorhinus brucus, Centroscymnus cryptacanthus* e *Somniosus microcephalus* [Squaliformes: Echinorhinidae, Squalidae], com a ampliação de suas distribuições. In: Proceedings of the VII Reunião do Grupo de Trabalho sobre Pesca e Pesquisa de Tubarões e Raias no Brasil; November 1994. Rio Grande-RS, Brazil. p. 66

[74] Nunan GW, Senna MLV. Tubarões [Selachii] coletados pelo navio oceanográfico "Thalassa" sobre a plataforma externa e talude continental do Brasil entre 11° e 22° S. In: Costa PAS, Olavo G, Martins AS, editors. Biodiversidade da fauna marinha profunda na costa central brasileira. Rio de Janeiro: Museu Nacional; 2007 [Série Livros n. 24]. pp. 163-183

[75] Akhilesh KV, Bineesh KK, White WT, Shanis CPR, Hashim M, Ganga U, Pillai NGK. Catch composition, reproductive biology and diet of the bramble shark *Echinorhinus brucus* (Squaliformes: Echinorhinidae) from the south-eastern Arabian Sea. Journal of Fish Biology. 2013;**83**:1112-1127 DOI:10.1111/jfb.12201

[76] Girard M, Du Buit, MH. Reproductive biology of two deep-water sharks from the British Isles, *Centroscymnus coelolepis* and *Centrophorus squamosus* [Chondrichthyes: Squalidae]. Journal of the Marine Biological Association UK. 1999;**79**:923-931

[77] Gordon JDM. Management considerations of deep-water shark fisheries. In: Shotton R, editor. Case Studies of the Management of Elasmobranch Fisheries. Rome-Italy: FAO Fisheries Technical Paper N° 378/2, part 2. 1999. pp. 774-818

[78] Clarke MW, Connoly PL, Bracken JJ. Catch, discarding, age estimation, growth and maturity of the squalid shark *Deania calceus* west and north of Ireland. Fisheries Research. 2002;**56**:139-153

[79] Guallart J, Serena F, Mancusi C, Casper BM, Burgess GH, Ebert DA, Clarke M, Stenberg C. *Centrophorus granulosus* [Internet]. 2006. In: IUCN 2006 Red List of Threatened Species. Available from: www.iucnredlist.org. [Accessed: 2017-03-10]

[80] Sadowsky V, Arfelli CA, Amorim AF. First record of broadbanded lanternshark, *Etmopterus gracilispinis* Kreft, 1968 [Squalidae], in the Brazilian waters. Boletim do Instituto de Pesca. 1986;**13**(2):1-4

[81] Soto JMR. Revisão da ocorrência de tubarões etmopteríneos na costa brasileira, com dois novos regis- tros de *Etmopterus bigelowi* Shirai and Tachikawa, 1993 (Chondrichthyes, Dalatiidae, Etmopterinae). Biotemas, 2000;**13**(2):167-178

[82] Rincon G, Lessa RPT. Evidências de sinonímia entre *E. pusillus* e *E. bigelowi* Shirai e Tachikawa, 1993, no nordeste do Brasil. In: Proceedings of the II Reunião da Sociedade Brasileira para o Estudo de Elasmobrânquios-SBEEL. November 2000. Santos: Instituto de Pesca; p. 73

[83] Gianeti MD, Vooren CM. Dieta de *Etmopterus bigelowi* Shirai e Tachikawa, 1993, no talude do Rio Grande do Sul. In: Proceedings of the V Reunião da Sociedade Brasileira para o Estudo de Elasmobrânquios-SBEEL. November 2006. Itajaí, Universidade do Vale do Itajaí. p. 35

[84] Gadig OBF, Sampaio CLS. Presença do gênero *Apristurus* [Chondrichthyes, Scyliorhinidae] no Brasil. In: Proceedings of the III Reunião da Sociedade Brasileira para o Estudo de Elasmobrânquios-SBEEL.November 2002. João Pessoa-PB. p. 42

[85] Gadig OBF, Gomes UL. Scyliorhinidae. In: Menezes NA, Buckup PA, Figueiredo JL, Moura RL,editors. Catálogo das espécies de peixes marinhos do Brasil. São Paulo-SP, Brazil: Museu de Zoologia, Universidade de São Paulo-USP; 2003. pp. 19-20

[86] Rincon G, Vaske-Júnior T, Vooren CM. Novo registro do cação lagarto *Galeus arae* na costa sul brasileira. Bol. da SBEEL. 1998; N°3, mar, p. 7

[87] Rincon G. *Galeus mincaronei*. In: IUCN 2006 Red List of Threatened Species [Internet]. 2006. Available from: www.iucnredlist.org [Accessed: 2017-03-10]

[88] Gomes UL, Peters GO, Carvalho MR, Gadig OBF. Anatomical investigation of the slender catshark *Schroederichthyes tenuis* Springer, 1966, with notes on intrageneric relationships [Chondrichthyes: Carcharhiniformes: Scyliorhinidae]. Zootaxa. 2006;**1119**:29-58

[89] Soares KDA, Gomes UL, Carvalho MR. Taxonomic review of catsharks of the *Scyliorhinus haeckelii* group, with the description of a new species (Chondrichthyes: Carcharhiniformes: Scyliorhinidae). Zootaxa. 2016;**4066**(5):501-534

[90] Springer S. A revision of the catsharks, Family Scyliorhinidae. NOAA Technical Report NMFS. 1979;**422**: p. 152

[91] Bacilieri S. Revisão taxonômica do gênero *Scyliorhinus* Blaiville, 1816 [Carcharhiniformes, Scyliorhinidae], nas regiões sudeste e sul do Brasil [thesis]. Rio Claro-SP, Brazil: Universidade Estadual Paulista-UNESP; 2005

[92] Soares KDA, Gadig OBF, Gomes UL. *Scyliorhinus ugoi*, a new species of catshark from Brazil (Chondrichthyes: Carcharhiniformes: Scyliorhinidae). Zootaxa. 2015;**3937**(2):347-361.DOI: http://dx.doi.org/10.11646/zootaxa.3937.2.6

[93] Vaz DFB, Carvalho MR. Morphological and taxonomic revision of species of *Squatina* from the Southwestern Atlantic Ocean (Chondrichthyes: Squatiniformes: Squatinidae). Zootaxa. 2013;**3695**:1-81

[94] Gadig OBF, Medina A, Bezerra MA, Furtado-Neto MAA. Dados sobre *Squatina dumeril* [Chondrichthyes: Squatinidae] no Brasil, com comentários taxonômicos do gênero *Squatina* em águas Brasileiras. Arquivos de Ciências do Mar. 1999;**32**:133-136

[95] McEachran JD, Compagno LJV. Results of the research cruises of FRV "Walter Herwig" to South America. LVI: A new species of skate from the Southwestern Atlantic, *Gurgesiella dorsalifera* sp. nov. [Chondrichthyes, Rajoidei]. Archiv für Fischereiwissenschaft. 1980; **31**:1-14

[96] Séret B, Andreata JV. Deep-sea fishes collected during cruise MD-55 off Brazil. Cybium. 1992;**6**(1):81-100

[97] Bernardes RA, Figueiredo JL, Rodrigues AR, Fischer LG, Vooren CM, Haimovici M, Rossi-Wongtschowski CLDB. Peixes da zona econômica exclusiva da região sudeste-sul do Brasil: Levantamento com armadilhas, pargueiras e rede de arrasto de fundo. São Paulo-SP, Brazil: Editora da Universidade de São Paulo-EDUSP; 2006

[98] Rincon-Filho G. Taxonomia, alimentação e reprodução da raia elétrica Benthobatis sp. (Torpediniformes: Narcinidae) no sul do Brasil [thesis]. Rio Grande-RS, Brazil:Fundação Universidade de Rio Grande; 1997

[99] Krefft G, Stehmann M. Ergebnisse der Forschungsreisen des FFS "Walter Herwig" nach Südamerika. XXXVI] Zweitere neue Rochenarten aus dem Südwest-atlantic: *Raja [Dipturus] leptocauda* und *Raja [Dipturus] trachyderma* spec. nov. [Chondrichthyes, Batoidei, Rajidae]. Archiv für Fischereiwissenschaft. 1975;**25**:77-97

[100] Soto JMR, Mincarone MM. *Dipturus diehli* sp. nov. A New Species of Skate [CHONDRICHTHYES, RAJIDAE] From Southern Brazil. Mare Magnum. 2001;**1**(1):3-6

[101] McEachran JD, Dunn KA. Phylogenetic analysis of skates, a morphologically conservative clade of Elasmobranchs [Chondrichthyes:Rajidae]. Copeia. 1998:271-290. DOI: 10.2307/1447424

[102] Visintin MR,Perez JAA. Vulnerabilidade de espécies capturadas pela pesca de emalhe-de-fundo no sudeste-sul do Brasil: produtividade-suscetibilidade (PSA). Boletim do Instituto de Pesca, São Paulo. 2013;**42**(1):119-133. DOI: 10.5007/1678-2305.2016v42 n1p119

[103] Barcelos BN. Ictiofauna marinha do Rio Grande do Sul. VII. *Torpedo nobiliana* Bonaparte, 1835.Arraia torpedo. Boletim Instituto de Ciencias Naturais Universidade do Rio Grande do Sul. 1963;**21**:15

[104] Sadowsky V. Primeiro registro da ocorrência do peixe elétrico, *Torpedo puelcha* Lahille, 1928 [Fam. Torpedinidae], nas águas do Brasil. Ciencia e Cultura. 1976;(supl.)28:421

[105] Figueiredo JL. Manual de peixes marinhos do sudeste do Brasil. I. Introdução.Cações, raias e quimeras. São Paulo-SP, Brazil: Museu de Zoologia da Universidade de São Paulo; 1977. p. 104

[106] Garcia ML, Menni RC, Jaureguizar AJ. *Torpedo puelcha* [Chondrichthyes, Torpedinoidea], an endemic species from the Argentinean Zoogeographic Province. Biogeographica. 2000;**76**:173-178

[107] Astarloa JMD, Figueroa DE, Lucifora L, Menni RC, Prenski BL, Chiaramonte G. New records of the Pacific sleeper shark, *Somniosus pacificus* [Chondrichthyes: Squalidae], from the southwestern Atlantic. Ichthyological Research. 1999;**46**(3):303-308

[108] Gadig OBF, Sampaio CLS. Ocorrência de *Somniosus pacificus*na costa central do Brasil [Squaliformes, Somniosidae]. In: Proceedings of the III Reunião da Sociedade Brasileira para o Estudo de Elasmobrânquios-SBEEL. João Pessoa-PB. 2002. p. 43

[109] Soto JMR, Consulim CEN, Souza-Filho MB. First record of Clark's fingerskate, *Dactylobatus clarkii* [Bigelow & Schroeder, 1958] [Chondrichthyes, Rajidae], in the South Atlantic. Mare Magnum. 2004;**2**(1-2):145-148

[110] Carvalho MR, Gomes UL, McEachran JD. First report of legskate genus *Cruriraja,* from the southwestern Atlantic Ocean [Chondrichthyes: Rajidae]. Cybium. 2006;**30**(4):343-354

[111] Menezes NA, Buckup PA, Figueiredo JL, Moura RL, editors. Catálogo das espécies de peixes marinhos do Brasil.São Paulo: Museu de Zoologia da Universidade de São Paulo; 2003. p. 160

[112] Hall MA, Alverson DL, Metuzals KI. By-catch; Problems and solutions. Marine Pollution Bulletin. 2000;**41**(1-6):204-219. DOI: 10.1016/S0025-326X(00)00111-9

[113] Kennelly SJ, Broadhurst MK. Fisherman and scientists solving by-catch problems: examples from Australia and possibilities for the Northeastern United States. In: Alaska Sea Grant College Program,editors.Solving By-catch: Considerations for Today and Tomorrow. Alaska Sea Grant College Program Report N° 96-03.Fairbanks: University of Alaska; 1996. pp. 121-128

[114] Denton EJ.Some adaptations of marine animals to physical conditions in the sea. In: Mauchline J, Nemoto T, editors. Marine Biology. Its Accomplishment and Future Prospect. Elsevier Publishing Company; 1991. p. 261

[115] Clarke MW, Kelly CJ, Connoly PL, Molloy JP. A life history approach to the assessment and management of deepwater fisheries in the Northeast Atlantic. Journal of Northwest Atlantic Fishery Science. 2003;**31**:401-411

[116] Last PR, Stevens JD. Sharks and Rays of Australia. SCIRO: Australia, Division of Fisheries; 1994. p. 513

[117] Perez JAA, Pezzuto PR, Andrade HA. Biomass assessment of the monkfish *Lophius gastrophysus* stock exploited by new deep-water fishery in southern Brazil. Fisheries Research. 2005;**72**:149-162. DOI:10.1016/j.fishres.2004.11.004

[118] Clarke MW. Records of deep-water chondrichthyan fish caught on long-line in the Rockall Trough. Journal of the Marine Biological Association of the UK. 2000;**80**:377-378

[119] Casey JM, Myers RA. Near extinction of a large widely distributed fish. Science. 1998;**281**:690-692. DOI: 10.1126/science.281.5377.690

[120] Dulvy NK, Sadovy Y, Reynolds JD. Extinction vulnerability in marine populations. Fish and Fisheries. 2003;**4**:25-64. DOI:10.1046/j.1467-2979.2003.00105.x

[121] Bernardes RA, Rossi-Wongstchowski CLDB, Wahrlich R, Vieira PC, dos Santos AP, Rodrigues AR. Prospecção pesqueira de recursos demersais com armadilhas e parguei-ras na Zona Econômica Exclusiva da Região Sudeste-Sul do Brasil. Série Documentos REVIZEE – Score Sul.São Paulo: Instituto Oceanográfico, USP; 2005. p. 112

[122] Sadovy Y, Vincent AJ. Ecological issues and the trades in live reef fishes. In: Sale PF, editor. Coral Reef Fishes: Dynamics and Diversity in a Complex Ecosystem. Academic Press: San Diego; 2002. pp. 391-420

[123] Large PA, Hammer C, Bergstad OA, Gordon JDM, Lorance P. Deep-water fisher-ies of the Northeast Atlantic: II assessment and management approaches. Journal of Northwest Atlantic Fishery Science. 2003;**31**:151-163

Chapter 6

A Review of the Mitogenomic Phylogeny of the Chondrichthyes

Divino Bruno da Cunha,
Luis Fernando da Silva Rodrigues-Filho and
João Bráullio de Luna Sales

Additional information is available at the end of the chapter

http://dx.doi.org/10.5772/intechopen.70028

Abstract

The phylogenetic analysis of the Chondrichthyes has been the subject of intense debate over the past two decades. The principal relationships within the group based on the analysis of morphological traits are inconsistent with the available molecular topologies, and the phylogeny of these animals is highly controversial, at all levels, ranging from superorders to families and even the genera within families. With the recent development of new generation sequencing (NGS), many phylogenies are now being inferred based on the complete genome of the species. In 2015 and 2016 alone, around 21 new elasmobranch genomes were made available in GenBank. In this context, the principal objective of the present study was to infer the phylogeny of the sharks and rays based on the complete mitochondrial genomes available in the literature. A total of 73 mitogenomes of chondrichthyan species were analyzed. The phylogenetic trees generated rejected the "Hypnosqualea" hypothesis and confirmed the monophyly of the Neoselachii and Batoidea as sister groups of the sharks. These mitogenomic analyses provided ampler and more complete insights into the relationships between the sharks and rays, in particular, the topologies obtained by the analyses revealed a number of incongruities in certain groups of sharks and rays, and the interrelationships between them.

Keywords: phylogeny, Chondrichthyes, mitogenomic, sharks and rays

1. Introduction

The Chondrichthyes (sharks, rays, and chimeras) are a diverse group of animals that occupy a key position in the phylogeny of the vertebrates, as one of their most ancient lineages

[1, 2]. Estimates of the diversification of the group based on DNA sequences are of the order of 300–460 million years ago. The class Chondrichthyes currently includes 14 orders and 60 families with approximately 1200 species [3–5]. The group of chondrichthyans with the greatest species diversity is the Batoidea (approximately 494 species), followed by the sharks, with around 500 species [4, 5]. The systematics of the Chondrichthyes have been the subject of considerable debate over the past two decades, and the phylogeny of these animals is still controversial, at all levels, ranging from superorders to families and even the genera within families [6].

Most of the hypotheses on the phylogenetic relationships among chondrichthyan species are based on morphological traits, in particular, the "Hypnosqualea" hypothesis [7, 8], which proposes that the batoids are derived sharks related intimately to the sawfish and angel sharks. However, most molecular phylogenies have rejected the "Hypnosqualea" hypothesis emphatically [3, 9–11].

A large number of morphological hypotheses have also been proposed to account for the arrangement of the shark orders, such as Ref. [12], which grouped the Squalomorphii and Squatinomorphii together, based on the synapomorphic nature of the orthostatic suspension of the maxilla within the ocular orbit, with the sharks of this clade being referred to as the "orthopedic" forms. In the case of the Galeomorphii and Squalimorphii, a number of studies, both molecular and morphological, have revealed a variety of controversial relationships within each group, and even the relationships among the orders vary between studies. One major example is the position of the Heterodontiformes, which has yet to be resolved [3, 7, 9, 10, 13, 14].

One other unresolved question is the phylogenetic position of the orders within the Galeomorphii, in which the available morphological inferences place Lamniformes as the sister order of the Carcharhiniformes [7, 15], although molecular studies have confirmed that the Orectolobiformes is the sister group of the Carcharhiniformes [13]. The topologies generated in other studies have nevertheless placed Lamniformes as the sister order of the Carcharhiniformes. On a more basal taxonomic level, there is even less agreement among the studies, and the taxonomic status of many Carcharhiniformes families is still unclear, such as the paraphyletism of the hammerhead sharks (Sphyrna) [14, 16, 17].

Another important question is the evidence of extensive paraphyly between the families of the Carcharhiniformes. Winchell et al. [18] concluded that the Scyliorhinidae is paraphyletic, a hypothesis confirmed by Refs. [12, 19]. Based on nuclear and mitochondrial sequences, these authors also proposed that the family Triakidae is paraphyletic, an arrangement confirmed by Ref. [20] based on a number of mitochondrial markers and the nuclear RAG1 gene, which contradicts the hypothesis of monophyly for the Triakidae. Similarly, the relationships among the members of the Batoidea remain unclear [15, 21–23]. In general, for example, four groups are recognized, the stingrays (Myliobatiformes), electric rays (Torpediniformes), sawfish (Pristiformes), skates, and guitarfish (Rajiformes), although the relationships among these groups are still undefined [21–25]. In addition, a number of new species of ray have been described recently, and the radiation patterns of this group have been better identified [25–29].

http://dx.doi.org/10.5772/intechopen.70028

The most recent molecular analyzed the complete mitochondrial genomes of 48 elasmobranch species, and rejected the "Hypnosqualea" hypothesis. The authors of this study recuperated the monophyly of the Lamnidae, with the families arranged as (Mitsukurinidae (Alopiidae, Megachasmidae) (Odontaspididae (Cetorhinidae (Lamnidae))) [6]. It is important to note that most chondrichthyan topologies are based on DNA sequences from only a few markers, such as a single nuclear or mitochondrial gene, or a combination of these two genes, and the sequences rarely exceed a length of 6 kb [3, 9–11], except for Ref. [6]. More recently, however, improvements in the speed and accuracy of new generation sequencing (NGS), and the reduction of laboratory costs, have provided a large number of molecular markers, amplifying considerably the analytical perspectives in the fields of phylogenetic and phylogenetic research, and the potential for the study of molecular ecology in a wide range of organisms [30–32]. In 2015 and 2016 alone, around 21 new elasmobranch genomes were made available in GenBank.

Mitochondrial markers are widely used in phylogenetic analyses in a diversity of organisms due to the relatively simple sequencing procedures and the high rates of nucleotide substitution [33]. However, the analysis of the complete mtDNA genome may provide a much richer source of genetic information for phylogenetic inferences in comparison with the more traditional approach, based on the analysis of single or multiple genes [32]. In this context, the principal objective of the present study was to provide phylogenetic inferences on the Chondrichthyes based on the complete mitochondrial genome, and a more comprehensive understanding of the relationships among the species of sharks and rays.

Order/species	Family	GenBank
Carcharhiniformes		
Carcharhinus leucas_G1	Carcharhinidae	NC023522
Carcharhinus leucas	Carcharhinidae	KJ210595
Carcharhinus macloti	Carcharhinidae	NC024862
Carcharhinus sorrah	Carcharhinidae	NC024055
Carcharhinus acronotus	Carcharhinidae	NC024596
Carcharhinus plumbeus	Carcharhinidae	KF801102
Carcharhinus falciformis	Carcharhinidae	KF801102
Carcharhinus obscurus	Carcharhinidae	NC020611
Carcharhinus melanopterus	Carcharhinidae	NC024284
Carcharhinus amblyrhyncoides	Carcharhinidae	NC023948
Carcharhinus longimanus	Carcharhinidae	KM434158.1
Carcharhinus brevipinna	Carcharhinidae	KM244770.1
Triaenodon obesus	Carcharhinidae	KJ748376.1
Prionace glauca	Carcharhinidae	NC022819
Glyphis garricki	Carcharhinidae	NC023361
Glyphis glyphis	Carcharhinidae	KF006312

Order/species	Family	GenBank
Galeocerdo cuvier cuvier	Carcharhinidae	NC022193
Scoliodon macrorhynchos	Carcharhinidae	JQ693102
Sphyrna zygaena	Sphyrnidae	KM489157
Sphyrna lewini	Sphyrnidae	NC022679
Sphyrna tiburo	Sphyrnidae	KM453976
Eusphyra blochii	Sphyrnidae	KU892590.1
Mustelus griseus	Triakidae	NC023527
Mustelus manazo	Triakidae	NC000890
Scyliorhinus canicula	Scyliorhinidae	NC022415
Lamniformes		
Carcharodon carcharias	Lamnidae	NC022691
Lamna ditropis	Lamnidae	NC024269
Isurus oxyrinchus	Lamnidae	NC022691
Isurus paucus	Lamnidae	NC024101
Cetorhinus maximus	Cetorhinidae	NC024101
Carcharias taurus	Odontaspididae	NC023266
Alopias pelagicus	Alopiidae	NC023520
Alopias superciliosus	Alopiidae	NC021443
Megachasma pelagios	Megachasmidae	NC021442
Mitsukurina owstoni	Megachasmidae	NC011825
Orectolobiformes		
Orectolobus japonicus	Orectolobidae	KF111729
Rhyncodon typus	Rhincodontidae	NC023455
Chiloscyllium griseum	Hemiscylliidae	NC017882
Chiloscyllium plagiosum	Hemiscylliidae	NC012570
Chiloscyllium punctatum	Hemiscylliidae	NC016686
Stegostoma fasciatum	Stegostomatidae	KU057952.1
Heterodontiformes		
Heterodontus francisci	Heterodontidae	NC003137
Heterodontus zebra	Heterodontidae	NC021615
Squatiniformes		
Squatina formosa	Squatinidae	NC025328
Squatina japonica	Squatinidae	NC024276
Squatina nebulosa	Squatinidae	NC025578
Squatina formosa	Squatinidae	NC025328

Order/species	Family	GenBank
Squaliformes		
Squalus acanthias	Squalidae	NC002012
Cirrhigaleus australis	Somniosidae	KJ128289
Squaliolus aliae	Dalatiidae	KU873080.1
Somniosus pacificus	Somniosidae	NC022734
Pristiophoriformes		
Pristiophorus japonicus	Pristiophoridae	NC_024102
Hexanchiformes		
Hexanchus griseus	Hexanchidae	KF894491
Myliobatiformes		
Gymnura poecilura	Gymnuridae	NC_024102
Dasyatis akajei	Dasyatidae	NC021132.1
Torpediformes		
Narcine entemedor	Narcinidae	KM386678
Rajiformes		
Rhinobatos schlegelii	Rhinobatidae	NC023951
Rhinobatos hynnicephalus	Rhinobatidae	NC022841
Rhina ancylostoma	Rhinobatidae	KU721837.1
Zearaja chilensis	Rajidae	KJ913073
Hongeo koreana	Rajidae	NC021963
Dipturus kwangtungensis	Rajidae	NC023505
Raja pulchra	Rajidae	NC025498
Raja rhina	Rajidae	KC914434
Okamejei hollandi	Rajidae	KP756687
Okamejei kenojei	Rajidae	NC007173
Atlantoraja castelnaui	Arhynchobatidae	NC025942
Pavoraja nitida	Arhynchobatidae	NC024599
Pristiformes		
Anoxypristis cuspidata	Pristidae	NC026307
Pristis clavata	Pristidae	KF381507
Pristis pectinata	Pristidae	NC027182
Chimaeriformes		
Callorhinchus milli	Callorhinchidae	NC014285
Chimaera monstrosa	Chimaeridae	AJ310140

Table 1. Complete mitochondrial genome of the elasmobranch species analyzed in the present study.

2. Materials and methods

All 73 mitogenomes analyzed in the present study were obtained from the GenBank public database (**Table 1**). The sequences were aligned automatically by Clustal and colleagues [34], run in the BioEdit sequence editor [35], which was used for visual inspection and possible corrections. The phylogenetic trees were rooted with the species *Callorhincus milli* and *Chimaera monstrosa*, which are considered to be most closely related to the sharks and rays, based on the results of previous studies [3, 10, 11]. The Bayesian inference was run in Mr. Bayes 3.0b4 [36], using the GTR + I + G model, which was selected by jModelTest 2 [37], based on the Bayesian information criterion (BIC) [38]. A Metropolis-coupled Markov chain Monte Carlo (MCMCMC) was executed with four chains run for 100,000,000 generations, using the default parameters. The quality of the run was verified in Tracer v1.6 [39]. The maximum likelihood tree was generated in PhyML 3.0 [40], using the GTR + I + G model, selected by jModelTest 2 [37], based on the Akaike information criterion (AIC), with the confidence interval being established by a bootstrap of 1000 replicates [41]. All the trees were visualized and edited in FigTree v.1.4.3 [42]. The distances among the taxa were calculated using NeighborNet, run in SplitsTree [43].

3. Results and discussion of the phylogenetic relationships

The phylogenetic analyses of the 73 shark and ray mitogenomes supports a basic division into four major groups, the Galeomorphii, Squalomorphii, Squatinomorphii, and Batoidea (**Figure 1**), as suggested in previous molecular studies, based on both mitochondrial and nuclear genes [3, 6, 9–11, 13]. The Bayesian and the maximum likelihood phylogenetic trees of the mitochondrial genomes had highly similar topologies, with the Batoidea being placed as the most basal group, sister group to all the others [3, 6, 11, 13]. The results of this analysis reject emphatically the "Hypnosqualea" hypothesis based on morphological data, which proposes that the Batoidea is part of the shark group, and represents a highly derived lineage of this group (see [7, 8, 44]). The molecular and morphological data are highly divergent with regard to this question.

The phylogenetic reconstruction obtained in the present study confirms the monophyly of the modern sharks (neoselachian), although another controversial question is the existence of the superorders proposed by Ref. [44], based on cladistic analyses of morphological data, which indicated the existence of four superorders—Galeomorphii, Squalomorphii, Squatinomorphii, and Batoidea. While the results of the present study confirm the existence of these four major groups, there are some differences in their arrangement (**Figure 1**) [6, 9, 11, 13].

The data support the monophyly of the order Squaliformes, with *Hexanchus griseus* (Hexanchiformes) at the base of the clade, followed by squaliformes as the sister taxon of the Squatiniformes and Pristiophoriformes. An important feature of the arrangement of the species within the Squatiniformes is the position of *Squatina nebulosa* as sister taxon of *S. formosa*, with *S. japonica* in a basal position in the clade. *Pristiophorus japonicus* is a sister group

http://dx.doi.org/10.5772/intechopen.70028

Figure 1. Bayesian inference and maximum likelihood generated from the complete mitochondrial genome of the Chondrichthyes. Only branches with bootstrap values below 100% are shown.

of this same clade, which highlights the paraphyletism of the two orders (Squatiniforme and Pristiophoriformes), an arrangement proposed by Ref. [13], who were nevertheless unable to define the position of *S. nebulosa*. The close relationship between the orders Squaliformes and Squatiniformes is consistent with [12], who used the projection of the cartilage of the upper mandible into the ocular orbit as a synapomorphic trait, with the sharks of this clade being denominated "orbitostylic" sharks. This arrangement confirms the findings of previous studies based on molecular markers [3, 6, 13, 18].

In the case of the Galeomorphii, it is important to note that the Heterodontiformes was identified as the most basal taxon of this superorder, in a clade supported by high probability values

(**Figure 1**). The Lamniformes is the sister group of Carcharhinifromes, which is placed as sister group of the Orectolobiformes with strong statistical support. The taxonomic position of the order Heterodontiforme within the Galeomorphii is considered to be extremely controversial [13]. Some studies have identified the Heterodontiformes as the most basal order, with Orectolobiformes being more closely related to the Lamniformes and Carcharhinifromes, a grouping supported by both morphological [7, 8] and molecular studies [3, 6, 9]. However, Ref. [18], using 5.8 kb of nuclear gene sequences (major and minor rRNA subunits), and Ref. [14], who analyzed mitochondrial data, indicated a closer relationship between the Orectolobiformes and Lamniformes [6, 10, 13].

The family-level phylogenetic inferences within the Carcharhinifromes reinforced the paraphyletism of this order, with the clade of the hammerhead sharks (Sphyrnidae) being placed as a sister group of the sharks (Carcharhinidae). *Scyliorhinus canicula* (Scyliorhinidae) was the most basal of the Carcharhiniformes, with *Mustelus manazo* (Triakidae) and *Galeocerdo cuvier* (Carcharhinidae) being placed close to all the other sharks, with high levels of statistical support (**Figure 1**). This arrangement is inconsistent with the results of Ref. [6], who used the complete mtDNA genome, and placed *Scoliodon macrorhincos* as a sister group of the sharks, with the sphyrnas as a sister group of this same clade. However, the arrangement observed here is in agreement with the studies of Refs. [3, 11, 13]. The most likely explanation for the lack of agreement between the results of the present study and those of Ref. [6] may be the difference in the number of species analyzed, given that an additional 24 mitogenomes were included in the present study. In addition, the pairs of species *Carcharhinus sorrah/C. macloti, C. amblyrhyncoides/C. leucas,* and *C. brevipania/Triaenodon obesus/C. acronotus* were involved in a polytomy. The polytomic pattern within Carcharhinidae is probably related to the low-levels of intrinsic genetic variability of sharks (**Figure 2**) [47]. The results of the present study are consistent with those of other molecular phylogenies [10, 13, 17, 24]. The analyses support the monophyly of the Lamniformes, with high probability values. The genus *Lamna* is the sister group of *Carcharodon* and *Isurus*, the most derived taxa of this group. This arrangement is supported by both the morphological data (Compagno 1990) and molecular inferences [6, 13, 17].

The monophyly of the rays was also supported by high probability values [3, 6, 9, 10, 16]. The analysis revealed the formation of four well-defined groups—Torpediformes, Rajiformes, Pristiformes, and Myliobatiformes—with branches supported by high bootstrap values (**Figure 1**). These results are consistent with previous studies based on the analysis of morphological and molecular data [21–25]. One other important finding of the present study was the arrangement of the orders Torpediniformes and Rajiformes close to the root of the tree (**Figure 2**). At the family level, two clades were observed, one containing the Narcinidae (the electric rays) as the most primitive taxon of the group, with the Rajidae (skates) and Arhynchobatidae grouped in the same clade, all on branches with high statistical support. The arrangement of the Narcinidae (Torpediniformes) is inconsistent with the morphological inferences, which assume that Pristiformes is the most basal order of the Batoidea [11, 15, 24]. However, the principal phylogenies based on mitochondrial and nuclear markers indicate that the Rajoidei is the sister group of all the other members of the Batoidea [3, 6, 16, 21, 22].

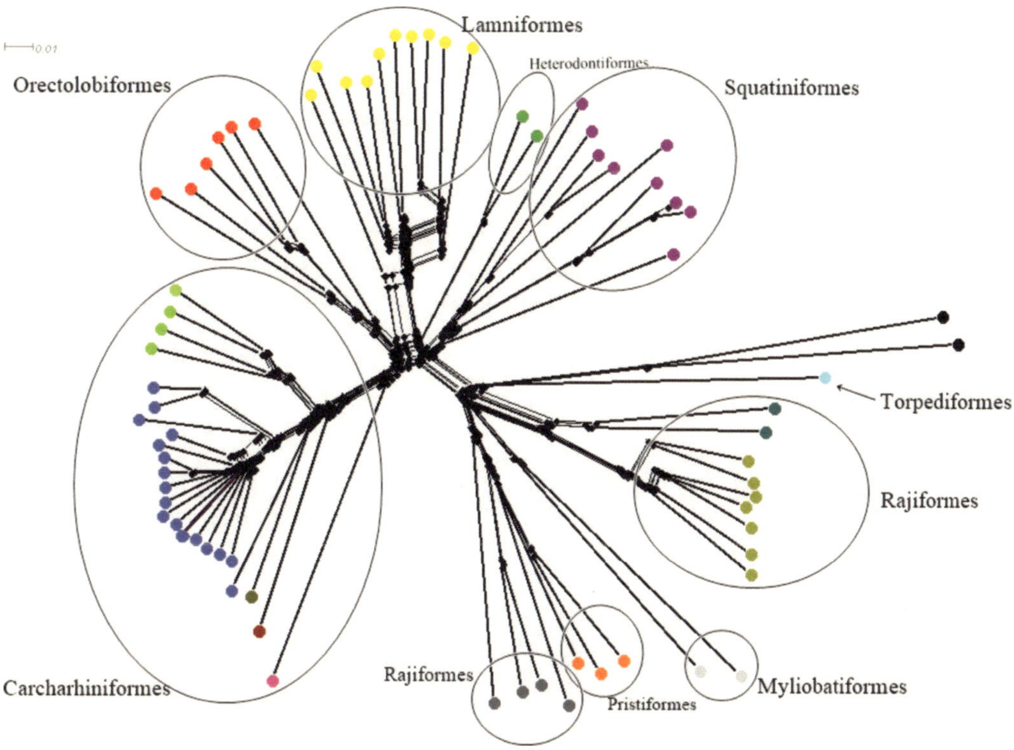

Figure 2. Genealogy showing the relationships between the orders of Chondrichthyes based on the complete mitochondrial genome of 73 species of sharks and rays. The two black circles on the right represent the out group.

The second clade includes *Zapteryx exasperata* (Rhinobatidae) as the most basal taxon, with *Rhinobatos hynnicephalus*/*R. schlegelii* as sister taxa. The family Rhinobatidae is placed as a sister group of the Pristidae, which is represented by *Anoxpristis cuspidata, Pristis clavata,* and *P. pectinata,* with all branches being supported by high probability values. The Myliobatidae is placed in this same clade as the sister group of the Rhinobatidae, an arrangement with high values in the Bayesian inference, and more moderate ones in the maximum likelihood analysis. The grouping of the sawfish (Pristiformes) with the guitarfish (Rhinobatidae) disagrees with the morphological phylogenies, which identifies the Pristiformes as the most primitive group of the Batoidea [23, 46]. However, this arrangement is supported by other molecular studies (see [3, 6, 16, 21]).

4. Conclusion

The analyses of the mitochondrial genome provided an ampler and more complete overview of the relationships within the Chondrichthyes, with the topologies highlighting a number of inconsistencies in some of the taxonomic groups of sharks and rays, principally in terms of the interrelationships among groups [13]. It is important to note that the phylogenetic relationships within the genera of these groups are still poorly understood, and that a mitogenomic phylogeny, including a much broader diversity of taxa, may provide more comprehensive insights into the relationships among the species of these organisms. Among other conclusions, the phylogenetic trees rejected the "Hypnosqualea" hypothesis and confirmed the monophyly of the Neoselachii, and Batoidea as the sister group of the sharks [16, 21], in agreement with most of the available molecular phylogenies. The Batoidea is a monophyletic group, in which the Torpediniformes and Rajiformes are the most ancestral orders, contradicting the morphological analyses, which identifies the Pristiformes as the most basal order. The phylogenetic trees supported conclusively the division of the Batoidea into four groups, the Torpediniformes, Rajiformes, Pristiformes, and Myliobatiformes [45].

Acknowledgements

We would like to thank the Bragança campus of the Federal University of Pará and the Evolution Laboratory. We are also grateful to the Northern Elasmobranchs study group.

Author details

Divino Bruno da Cunha[1*], Luis Fernando da Silva Rodrigues-Filho[2] and João Bráullio de Luna Sales[1,3]

*Address all correspondence to: divinobruno@yahoo.com.br

1 Institute for Coastal Studies (IECOS), Federal University of Pará, Bragança, Pará, Brazil

2 Biological Sciences Faculty, Federal Rural University of the Amazon, Capanema, PA, Brazil

3 Faculdade de Ciencias Naturais (FACIN), Federal University of Pará, Breves, PA, Brazil

References

[1] Renz AJ, Meyer A, Kuraku S. Revealing less derived nature of cartilaginous fish genomes with their evolutionary time scale inferred with nuclear genes. PLoS ONE. 2013;**8**:e66400. DOI: 10.1371/journal.pone.0066400

[2] Nelson JM. Fishes of the World. New York: Wiley; 1994

[3] Heinicke MP, Naylor GJP, Hedges SB. Cartilaginous fishes (Chondrichthyes). In: Hedges SB, Kumar S, editors. The Timetree of Life. New York: Oxford University Press; 2009. p. 320

[4] Weigmann S. Annotated checklist of the living sharks, batoids and chimaeras (Chondrichthyes) of the world, with a focus on biogeographical diversity. Journal of Fish Biology. 2016;**88**:837-1037. DOI: 10.1111/jfb.12874

[5] Gemaque R, Monteiro ILP, Gomes F, Sodré D, Sampaio I, Sales JBL, Rodrigues Filho LFS. Why implement measures to conserve the diversity of Elasmobranchs? The case of the northern coast of Brazil. Revista da Biologia. 2017;**17**:1-6. DOI: 10.7594/revbio.17.02.01

[6] Diaz-Jaimes P, Uribe-Alcocer M, Hinojosa-Alvarez S, Sandoval-Laurrabaquio N, Adams DH, García De León FJ. The complete mitochondrial DNA of the bull shark (*Carcharhinus leucas*). Mitochondrial DNA. 2014;**27**(1):717-8. DOI: 10.3109/19401736.2014.913157

[7] de Carvalho MR, Maisey JG. Phylogenetic relationships of the late jurassic shark protospinax Woodward 1919 (Chondrichthyes: Elasmobranchii). In: Arratia G, Viohl G, editors. Mesozoic Fishes: Systematics and Paleoecology. Munich: Verlag Dr Friedrich Pfeil; 1996. pp. 9-46

[8] Shirai S: Phylogenetic interrelationships of neoselachians (Chondrichthyes: Euselachii). In: Stiassny MLJ, Parenti LR, Johnson GD, editors. Interrelationships of Fishes. San Diego: Academic Press; 1996. pp. 9-34

[9] Naylor GJP, Ryburn JA, Fedrigo O, López JA. Phylogenetic relationships among the major lineages of modern elasmobranchs. In: Hamlett WC, Jamieson BGM, editors. Reproductive Biology and Phylogeny 3. Science Publishers; 2005. pp. 1-25

[10] Human BA, Owen EP, Compagno LJV, Harley EH. Testing morphologically based phylogenetic theories within the cartilaginous fishes with molecular data, with special reference to the catshark family (Chondrichthyes; Scyliorhinidae) and the interrelationships within them. Molecular Phylogenetics and Evolution. 2006;**39**:384-391

[11] Gkafas GA, Megalofonou P, Batzakas G, Apostolidis AP, Exadactylos A. Molecular phylogenetic convergence within Elasmobranchii revealed by cytochrome oxidase subunits. Biochemical Systematics and Ecology. 2015;**61**:510-515. DOI: 10.1016/j.bse.2015.07.025

[12] Maisey JG. Higher elasmobranch phylogeny and biostratigraphy. Zoological Journal of the Linnean Society. 1984;**82**:33-54

[13] Vélez-Zuazo X, Agnarsson I. Shark tales: A molecular species-level phylogeny of sharks (Selachimorpha, Chondrichthyes. Molecular Phylogenetics and Evolution. 2011; **58**:207-217

[14] Douady CJ, Dosay M, Shivji MS, Stanhope MJ. Molecular phylogenetic evidence refuting the hypothesis of Batoidea (rays and skates) as derived sharks. Molecular Phylogenetics and Evolution. 2003;**26**:215-221

[15] Compagno LJV. Interrelationships of living elasmobranchs. Zoological Journal of the Linnean Society. 1973;**53**:15-61

[16] Naylor GJP, Caira JN, Jensen K, Rosana KAM, White WT, Last PR. A sequence based approach to the identification of shark and ray species and its implications for global elasmobranch diversity and parasitology. Bulletin of the American Museum of Natural History. 2012;**367**:1-262

[17] Lim DD, Motta P, Mara K, Martin AP. Phylogeny of hammerhead sharks (family Sphyrnidae) inferred from mitochondrial and nuclear genes. Molecular Phylogenetics and Evolution. 2010;**55**:572-579

[18] Winchell CJ, Martin AP, Mallatt J. Phylogeny of elasmobranchs based on LSU and SSU ribosomal RNA genes. Molecular Phylogenetics and Evolution. 2004;**31**:214-224

[19] Iglésias SP, Lecointre G, Sellos DY. Extensive paraphylies within sharks of the order Carcharhiniformes inferred from nuclear and mitochondrial genes. Molecular Phylogenetic Evolution. 2005;**34**:569-583

[20] Lopez JA, Ryburn JA, Fedrigo O, Naylor GJ. Phylogeny of sharks of the family Triakidae (Carcharhiniformes) and its implications for the evolution of carcharhiniform placental viviparity. Molecular Phylogenetic Evolution. 2006;**40**:5060

[21] Aschliman NC, Nishida M, Miya M, Inoue JG, Rosana KM, Naylor GJP. Body plan convergence in the evolution of skates and rays (Chondrichthyes: Batoidea). Molecular Phylogenetics and Evolution. 2012;**63**:28-42. DOI: 10.1016/j.ympev.2011.12.012

[22] Pavan-Kumar A, Gireesh-Babu P, Babu PP, Jaiswar AK, Hari Krishna V, Prasasd KP. Molecular phylogeny of elasmobranchs inferred from mitochondrial and nuclear markers. Molecular Biology Reports. 2013;**41**:447-457. DOI: 10.1007/s11033-013-2879-6

[23] McEachran JD, Aschliman N. Phylogeny of batoidea. In: Carrier JC, Musick JA, Heithaus MR, editors. Biology of Sharks and Their Relatives. Boca Raton, FL: CRC Press; 2004. pp. 79-113

[24] Nishida K. Phylogeny of the suborder Myliobatoidei. Mem. Fac. Fish Hokkaido Univ. 1990;**37**:1-108

[25] Naylor GJP, Caira JN, Jensen K, Rosana KAM, Straube N, Lakner C. Elasmobranch phylogeny: A mitochondrial estimate based on 595 species. In: Carrier JC, Musick JA, Heithaus MR, editors. Biology of Sharks and Their Relatives. 2nd ed. Boca Raton: CRC Press; 2012. pp. 31-56

[26] Last PR, Naylor GJP, Manjaji-Matsumoto M. A revised classification of the family Dasyatidae (Chondrychthyes: Myliobatiformes) based on new morphological and molecular insights. Zootaxa. 2016;**4139**:345-368

[27] Last PR, Henderson AC, Naylor GJP. Acroteriobatus omanesis (Batoidea: Rhinobatidae), a new guitarfish from the Gulf of Oman. Zootaxa. 2016;**4144**:276-286

[28] Last PR, Kyne PM, Compagno LJV. A new species of wedgefish Rhynchobatus cooki (Rhinopristiformes, Rhinidae) from the Indo-West Pacific. Zootaxa. 2016;**4139**:233-247

[29] Puckridge M, Last PR, White WT, Andreakis N. Phylogeography of the indo-west pacific maskrays (Dasyatis, Neotrygon): A complex example of chondrichthyan radiation in the Cenozoic. Ecology and Evolution. 2013;**3**:217-232

[30] Avise JC. Molecular Markers, Natural History and Evolution. Sunderland MA: Chapman Hall; 1994. p. 511

[31] Templeton AR. The role of molecular genetics in speciation studies. In: Schierwater B, Streit B, Wagner GP, editors. Molecular Ecology and Evolution: Approaches and Applications. Basel: Birkhäuser Verlag; 1994. pp. 455-475

[32] Qin J, Zhang Y, Zhou X, Kong X, Wei S, Ward RD, Zhang A. Mitochondrial phylogenomics and genetic relationships of closely related pine moth (Lasiocampidae: Dendrolimus) species in China, using whole mitochondrial genomes. BMC Genomics. 2015;**16**:428-439

[33] Boore JL. Animal mitochondrial genomes. Nucleic Acids Research. 1999;**27**:1767-1780

[34] Thompson JD, Higgins DG, Gibson TJ, Clustal W. Improving the sensitivity of progressive multiple sequence alignment through sequence weighting, position-specific gap penalties and weight matrix choice. Nucleic Acids Research. 1994;**22**:4673-4680

[35] Hall TA. BioEdit: A user-friendly biological sequence alignment editor and analysis program for Windows 95/98/NT. Nucleic Acids Symposium Series. 1999;**41**:95-98

[36] Ronquist F, Huelsenbeck JP. MRBAYES 3: Bayesian phylogenetic inference under mixed models. Bioinformatics. 2003;**19**:1572-1574

[37] Darriba D, Taboada GL, Doallo R, Posada D. jModelTest 2: More models, new heuristics and parallel computing. Nature Methods. 2012;**9**:772-772

[38] Posada D, Buckley TR. Model selection and model averaging in phylogenetics: Advantages of akaike information criterion and bayesian approaches over likelihood ratio tests. Systematic Biology. 2004;**53**:793-808

[39] Rambaut A, Drummond AJ. Tracer version 1.5 [computer program] [Internet]. 2009. Available from: http://beast.bio.ed.ac.uk

[40] Guindon S, Dufayard J-F, Lefort V, Anisimova M, Hordijk W. New algorithms and methods to estimate maximum-likelihood phylogenies: Assessing the performance of PhyML 3.0. Systematic Biology. 2010;**59**:307-321

[41] Felsenstein J. Confidence limits on phylogenies: An approach using the bootstrap. Evolution. 1985;**39**:783-791

[42] Rambaut A. FigTree v1.4 [Internet]. 2013. Available from: http://tree.bio.ed.ac.uk/software/figtree/

[43] Huson DH, Bryant D. Application of phylogenetic networks in evolutionary studies. Molecular Biology and Evolution. 2006;**23**:254-267

[44] Compagno LJV. Phylogenetic relationships of living sharks and rays. American Zoologist. 1977;**17**:303-322

[45] White WT, Naylor GJP. Resurrection of the family Aetobatidae (Myliobatiformes) for the pelagic eagle rays, genus Aetobatus. Zootaxa. 2016;**4139**:435-438

[46] McEachran JD, Dunn KA, Miyake T. Interrelationships within the batoid fishes (Chondrichthyes: Batoidea). In: Stiassney MLJ, Parenti LR, Johnson GD, editors. Interrelationship of Fishes. New York: Academic Press; 1996. pp. 63-84

[47] Rodrigues-Filho LFS, Rocha TC, Rêgo PS, Schneider H, Sampaio I, Vallinoto M. Identification and phylogenetic inferences on stocks of sharks affected by the fishing industry off the Northern coast of Brazil. Genetics and Molecular Biology. 2009;**32**:405-413